T0295267

Proceedings of the

Eighth Meeting on CPT and Lorentz Symmetry

Proceedings of the

Eighth Meeting on CPT and Lorentz Symmetry

Indiana University, Bloomington, USA   12 – 16 May 2019

Editor

Ralf Lehnert

Indiana University, USA

 World Scientific

NEW JERSEY · LONDON · SINGAPORE · BEIJING · SHANGHAI · HONG KONG · TAIPEI · CHENNAI · TOKYO

Published by

World Scientific Publishing Co. Pte. Ltd.

5 Toh Tuck Link, Singapore 596224

USA office: 27 Warren Street, Suite 401-402, Hackensack, NJ 07601

UK office: 57 Shelton Street, Covent Garden, London WC2H 9HE

Library of Congress Control Number: 2020008815

British Library Cataloguing-in-Publication Data
A catalogue record for this book is available from the British Library.

CPT AND LORENTZ SYMMETRY
Proceedings of the Eighth Meeting on CPT and Lorentz Symmetry

ISBN 978-981-121-397-7 (hardcover)
ISBN 978-981-121-398-4 (ebook for institutions)
ISBN 978-981-121-399-1 (ebook for individuals)

For any available supplementary material, please visit
https://www.worldscientific.com/worldscibooks/10.1142/11655#t=suppl

Preface

The *Eighth Meeting on CPT and Lorentz Symmetry*, CPT'19, was held in the Physics Department at Indiana University, Bloomington from Sunday May 12 to Thursday May 16, 2019. Following the focus and format of seven previous conferences in this triennial series, the meeting covered current results and trends in experimental and theoretical research involving departures from spacetime symmetries.

The contributions contained in this volume provide a cross section of the individual topics presented at CPT'19. On the experimental side, they include studies at accelerators and colliders involving neutral mesons, the top quark, and the photon; astrophysical observations of cosmic rays, photons, and gravitational waves; gravity tests in the laboratory and in space; spectroscopic investigations of ions, atoms, molecules, and exotic atoms; measurements involving spin; matter–antimatter comparisons; lasers and masers; studies with cavities, oscillators, and resonators; and neutrino mixing, propagation, and endpoint measurements. The theoretical and phenomenological contributions comprise the classification and identification of effects in the Standard Model, General Relativity, and beyond; origins and mechanisms for spacetime-symmetry violations; analyses in field theory, gravitation, and particle physics; applications to condensed-matter systems; and investigations of the mathematical foundations and Finsler geometry. The reports are arranged according to the schedule of talks followed by the posters.

This meeting would not have been possible without the close cooperation with Alan Kostelecký. I am also tremendously grateful for the logistic support before and during the event by numerous staff members and colleagues including Kathy Hirons, Ágnes Roberts, Jeffery Coulter, Ben Edwards, Zonghao Li, Navin McGinnis, Nathan Sherill, and Rui Xu. Financial assistance for the meeting by the Indiana University Center for Spacetime Symmetries (IUCSS), the Indiana University Office of the Vice Provost for Research, and the United States Department of Energy is gratefully acknowledged.

Ralf Lehnert
November 2019

Contents

Rabi Experiments on the σ and π Hyperfine Transitions in Hydrogen and Status of ASACUSA's Antihydrogen Program

M.C. Simon

Stefan Meyer Institute, Austrian Academy of Sciences,
Boltzmanngasse 3, 1090 Wien, Austria

On behalf of the ASACUSA Collaboration[*]

We report on the status of the in-beam hyperfine-structure measurements on ground-state antihydrogen by ASACUSA and on recent results obtained in supporting measurements from hydrogen. The σ_1 and π_1 transitions can now be investigated, which is beneficial from both theoretical and experimental perspectives. We discuss systematic effects from resonance interference originating from the chosen field geometries in the interaction region, and how their impact can be managed by appropriate data-taking or design concepts.

1. Status of ASACUSA's antihydrogen program

ASACUSA (Atomic Spectroscopy And Collisions Using Slow Antiprotons) is one of several collaborations studying antimatter at the antiproton decelerator at CERN. The majority of experiments in this area compare antimatter properties to those of their matter counterparts for a precise test of CPT symmetry. Recent results on the magnetic moment of the antiproton[1] by BASE as well as the 1S–2S, 1S–2P, and hyperfine transitions of antihydrogen ($\overline{\text{H}}$) by ALPHA[2] demonstrate the rapid progress in this field. All existing measurements confirm CPT symmetry within their respective uncertainties. Thus, the quest for ever higher precision tests by the antiproton-decelerator community continues. In this spirit, ASACUSA is working toward a hyperfine-splitting determination based on Rabi spectroscopy. The SME coefficients, which can be tested or constrained by this specific experiment have been discussed in the proceedings of the previous meeting in this series.[3] The approach followed by ASACUSA requires the formation of a beam of $\overline{\text{H}}$ and offers the advantage that the actual measurement, i.e., the interaction with microwaves, takes place in a well-controlled

---

[*]http://asacusa.web.cern.ch/ASACUSA/asacusaweb/members_files/members.shtml

environment far away from the strong fields and field gradients of the trap for $\overline{\text{H}}$ formation.[4,5] The observation of extracted antiatoms had been reported in Ref. 6, and later this beam had its quantum-state distribution characterized in Ref. 7. Ground-state $\overline{\text{H}}$ has clearly been observed. However, a rate increase is still needed before spectroscopic measurements with reasonable acquisition times can commence.

Due to the current long shutdown at CERN, antiproton physics is on hold until 2021. Then, the Extra-Low ENergy Antiproton ring (ELENA) will be in operation and the $\overline{\text{H}}$ experiment of ASACUSA will receive a dedicated beamline. Therefore, the multi-trap setup does not have to be removed from the zone anymore in order to make space for other antimatter studies pursued by ASACUSA. Currently, the setup is being installed at its final position and matter studies (i.e., mixing of protons and electrons) are planned to continue optimizing the formation process during the shutdown.

2. Transitions within the hyperfine sublevels

The two transitions that occur between a low-field seeking triplet state ($F = 1$) and the singlet state ($F = 0$) are called σ_1 and π_1 transitions (see Table 1). Those are accessible in a Rabi-type experiment and approach the value of interest at zero field. In the interaction volume, one has to provide both an oscillating magnetic field B_{osc} at 1.42 GHz to stimulate hyperfine transitions, and an external static magnetic field B_{ext} to control the Zeeman shift. A key difference between the two transitions is that they need different relative orientations of B_{osc} and B_{ext}. The component of B_{osc} aligned parallel to B_{ext} stimulates the σ_1 transition, and the orthogonal one the π_1 transition. In the ASACUSA setup, these fields are provided by a cavity of strip-line geometry with a large acceptance and surrounding coils.

Table 1: Properties of the hyperfine sublevels of ground-state $\overline{\text{H}}$.

State: $\lvert F, M_F \rangle$	Zeeman Shift		Stern–Gerlach	σ_1	π_1
$\lvert\, 1, -1\, \rangle$	$1^{\text{st}} O$	(linear)	low-field seeker		initial
$\lvert\, 1,\ \ 0\, \rangle$	$2^{\text{nd}} O$	(hyperbolic)	low-field seeker	initial	
$\lvert\, 1,\ \ 1\, \rangle$	$1^{\text{st}} O$	(linear)	high-field seeker		
$\lvert\, 0,\ \ 0\, \rangle$	$2^{\text{nd}} O$	(hyperbolic)	high-field seeker	final	final

In the low-field regime, the σ_1 transition has a weak magnetic-field sensitivity making it more robust against systematic effects. However, it is also insensitive to SME coefficients,[8] which relates to the fact that $\Delta M_F = 0$.

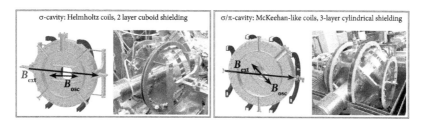

Fig. 1: ASACUSA's devices for providing an oscillating B_{osc} and external static magnetic field B_{ext} in the interaction region of the Rabi experiment using microwave cavities and coil configurations.

In contrast, the π_1 transition probes the nonrelativistic spherical SME co-effients $g_{w(2q)10}^{\text{NR(0B)}}$, $g_{w(2q)10}^{\text{NR(1B)}}$ (CPT odd) and $H_{w(2q)10}^{\text{NR(0B)}}$, $H_{w(2q)10}^{\text{NR(1B)}}$ (CPT even).[3,8] Control over systematic effects is a much stronger experimental challenge due to the first-order shift with B_{ext} of 14 Hz/nT.

Helmholtz coils produce a sufficiently homogeneous static field to investigate the σ_1 transition $(\sigma_{|B|}/|\bar{B}| < 5\%)$. The cavity providing the oscillating field is placed within those coils in such a way that the required parallel alignment is achieved, as shown in the left panel of Fig. 1. For more details and a description of the entire H-beam setup for commissioning the hyperfine spectrometer, see Ref. 10. A determination of the H hyperfine structure with 2.7 ppm precision in agreement with the literature value was accomplished[9] by measuring the σ_1 transition at various static-field values and extrapolating to zero field. However, as mentioned before the σ_1 transition is not the first choice for a CPT test, which motivated upgrades of the interaction apparatus for investigations of the π_1 transition.

We designed McKeehan-like coils[11] to provide a more homogeneous magnetic field $(\sigma_{|B|}/|\bar{B}| < 0.1\%)$. The cavity can be rotated in steps of 45° within the coil arrangement thereby providing flexible alignment of B_{osc} to B_{ext}. A 45° alignment enables simultaneous access to both transitions. The upgraded device is shown in the right panel of Fig. 1.

3. Interference effect and future measurements and devices

At small B_{ext}, the separation of the σ_1 and π_1 transitions becomes comparable to the line widths, which is on the order of 10 kHz as the interaction time is typically 100 μs (beam velocity ∼ 1 km/s, cavity length ∼ 100 mm). For example, at 4.6 μT the separation decreases to 65 kHz, and the interference leads to asymmetric line shapes and systematic shifts of the extracted central frequencies, if one applies the symmetric fit function for a two-level system in this regime. Currently, we apply corrections for the effect, and the development of a fit procedure based on the complete four-level system of the ground-state hyperfine states is under consideration. On the other hand, the interference effect and resulting systematic shifts can be avoided with purely parallel or orthogonal alignment of the fields. This clean solution will be realized in a Ramsey apparatus that is currently in its final design phase.[13] In principle, the respective alignment can also be adjusted in steps of $45°$ with the present Rabi apparatus. However, a change requires breaking the vacuum. In this design, the advantageous opportunity for interleaved measurements of π_1 and σ_1 transitions is tied to the alignment of the fields that also gives rise to the interference effect. By operating at sufficiently high B_{ext} (e.g., $23/69/115\,\mu$T as in Ref. 12) the systematic shifts and related uncertainties can easily be kept below $\sim 10\,$ppb, an acceptable level for the anticipated first-stage results on $\overline{\text{H}}$. A measurement of a single pair of a π_1 and σ_1 transitions in this regime of B_{ext} provides a reliable way to determine the zero-field splitting from a calculation,[12] i.e., without extrapolating to zero field by a fit. Thus, the present apparatus is well suited for a first Rabi-type SME-sensitive $\overline{\text{H}}$ hyperfine-structure measurement.

Acknowledgments

This work was funded by the European Research Council (grant No. 291242), the Austrian Ministry of Science and Research, and the Austrian Science Fund (FWF) through DKPI (W1252).

References

1. C. Smorra *et al.*, Nature **550**, 317 (2017).
2. T. Friesen, these proceedings.
3. E. Widmann *et al.*, in V.A. Kostelecký, ed., *Proceedings of the Seventh Meeting on CPT and Lorentz Symmetry*, World Scientific, Singapore, 2017.
4. E. Widmann *et al.*, Hyperfine Int. **215**, 1 (2013).
5. A.V. Mohri and Y. Yamazaki, Europhys. Lett. **63**, 207 (2003).
6. N. Kuroda *et al.*, Nat. Commun. **5**, 3089 (2014).
7. M. Malbrunot *et al.*, Phil. Trans. A **376**, 2116 (2018).
8. V.A. Kostelecký and A.J. Vargas, Phys. Rev. D **92**, 056002 (2015).
9. M. Diermaier *et al.*, Nat. Commun. **8**, 15749 (2017).
10. M. Malbrunot *et al.*, Nucl. Instrum. Meth. A **935**, 110 (2019).
11. L.W. McKeehan, Rev. Sci. Instrum. **7**, 178 (1936).
12. E. Widmann *et al.*, Hyperfine Int. **240**, 1 (2018).
13. A. Nanda *et al.*, these proceedings.

Lorentz Violation in Chiral Perturbation Theory

Brett Altschul

*Department of Physics and Astronomy, University of South Carolina,
Columbia, SC 29208, USA*

Most Lorentz-violating effects involving hadrons would be caused by Lorentz violation existing at the quark and gluon level. However, it can be difficult to relate experimental bounds on hadron-level Lorentz violation to the underlying parameters of the fundamental quark and gluon lagrangian. Chiral perturbation theory allows for a mapping between the types of Lorentz violation existing at these two levels and also between the types of Lorentz violation for different types of hadrons.

Over the last two decades, interest in Lorentz violation has grown because of the development of an effective field theory known as the Standard-Model Extension (SME), which can be used to describe all forms of Lorentz violation that may exist in a quantum field theory built out of the Standard-Model fields.[1] However, understanding of the SME is still far from complete. The SME is formulated as a relativistic field theory in terms of the fundamental quark, lepton, gauge, and Higgs fields of the Standard Model, which makes the relationships between the parameters in the fundamental SME lagrangian and experimental observables rather complicated, because at low energies, the Standard Model's strongly interacting degrees of freedom are not quarks and gluons, but actually composite hadrons.

Using chiral perturbation theory (χPT), it is possible to provide a quantitative mapping between the parton- and hadron-level parametrizations of Lorentz violation. χPT is an effective field theory that works by considering all terms at the hadronic level that are allowed by the symmetries of the underlying theory. For pure Quantum Chromodynamics (QCD), these symmetries include the discrete operations C, P, and T. In addition, QCD possesses an accidental chiral symmetry in the limit of vanishing quark masses. Physically, the masses of the u and d quarks are much smaller than the masses of typical hadrons, which means that setting $m_u = m_d = 0$ is a reasonable starting point for the construction of the effective lagrangian. In two-flavor QCD, the $SU(2)_L \times SU(2)_R$ chiral symmetry is spontaneously broken, with the pions as the corresponding (pseudo-)Goldstone bosons.

The pion fields can be collected in the $SU(2)$ matrix

$$U(x) = \exp\left[i\frac{\phi(x)}{F}\right], \tag{1}$$

where $\phi = \sum \phi_a \tau_a$ in terms of the $SU(2)$ generators, and F is the pion decay constant. The Lorentz-invariant term containing the kinetic Lagrange density for the pions is

$$\mathcal{L}_\pi = \frac{F^2}{4} \text{Tr}(\partial_\mu U \partial^\mu U^\dagger), \tag{2}$$

with the trace Tr being taken over flavor space. This can be generalized to include Lorentz violation. If the quark sector includes c-type (or d-type) operators of the form

$$\mathcal{L}_{\text{light quarks}}^{\text{CPT-even}} = i\bar{Q}_L C_{L\mu\nu} \gamma^\mu D^\nu Q_L + i\bar{Q}_R C_{R\mu\nu} \gamma^\mu D^\nu Q_R, \tag{3}$$

where $Q_{L/R} = [u_{L/R}, d_{L/R}]^T$ are the quark isodoublet, and the couplings are collected in the (μ, ν)-symmetric matrices

$$C_{L/R}^{\mu\nu} = \begin{bmatrix} c_{u_{L/R}}^{\mu\nu} & 0 \\ 0 & c_{d_{L/R}}^{\mu\nu} \end{bmatrix}, \tag{4}$$

then the most general nonvanishing, Lorentz-violating (but chirally symmetric) term that can be added to the pion Lagrange density at leading order is[2,3]

$$\mathcal{L}_\pi' = \beta^{(1)} \frac{F^2}{4} \left({}^1C_{R\mu\nu} + {}^1C_{L\mu\nu}\right) \text{Tr}[(\partial^\mu U)^\dagger \partial^\nu U]. \tag{5}$$

The Lorentz-violating contribution must take this form at leading order so that it is invariant under a chiral transformation of $U(x)$,

$$U(x) \to U'(x) = RU(x)L^\dagger, \tag{6}$$

where $(L, R) \in SU(2)_L \times SU(2)_R$.

The quantity $\beta^{(1)}$ is an initially undetermined "low-energy constant." Its value depends on the details of the nonperturbative dynamics that underlie the hadron structure. However, it can be estimated to be of $\mathcal{O}(1)$. Moreover, once its value is established through an analysis of one specific kind of process, the same constant value will appear in other interaction processes.

This is important, because while $\mathcal{L}_\pi + \mathcal{L}'_\pi$ contains the kinetic terms for the pions, it also contains higher-order interaction vertices as well. Expanding $U(x)$ in terms of the pion fields shows that the Lagrange density in Eq. (5) not only contains corrections to the pion propagator, but also induces new multi-pion interactions. The two-pion portion of $\mathcal{L}'_\pi$ is

$$\mathcal{L}'_{2\pi} = \frac{\beta^{(1)}}{4}(c^{\mu\nu}_{u_L} + c^{\mu\nu}_{d_L} + c^{\mu\nu}_{u_R} + c^{\mu\nu}_{d_L})\partial_\mu\phi_a\partial_\nu\phi_a. \tag{7}$$

This takes a standard form for Lorentz violation with a spinless field; conventionally, this type of Lorentz violation is described by a coefficient $k^{\mu\nu}$. Written in terms of the physical fields, Eq. (7) is

$$\mathcal{L}'_{2\pi} = \frac{\beta^{(1)}}{4}(c^{\mu\nu}_{u_L} + c^{\mu\nu}_{d_L} + c^{\mu\nu}_{u_R} + c^{\mu\nu}_{d_L})(\partial_\mu\pi^+\partial_\nu\pi^- + \partial_\mu\pi^-\partial_\nu\pi^+ + \partial_\mu\pi^0\partial_\nu\pi^0). \tag{8}$$

This has the form expected for a minimal SME term involving a spin-0 field. In the general Lagrange density

$$\mathcal{L}_{\text{spin-0}} = \frac{1}{2}\partial^\mu\phi_a\partial_\mu\phi_a + \frac{1}{2}k^{\mu\nu}\partial_\mu\phi_a\partial_\nu\phi_a - \frac{m^2}{2}\phi_a\phi_a, \tag{9}$$

the tensor $k^{\mu\nu}$ modifies the equations of motion—or, equivalently, the energy–momentum relation for freely propagating particles. These propagation modifications can have many observable physical consequences. When particles have modified energy–momentum relations, which do not have the standard relativistic forms, there may be upper and lower thresholds for various decay and emission processes. For example, with an appropriate choice of parameters, the decay of photons into charged particle–antiparticle pairs (such as $\gamma \to \pi^+ + \pi^-$) may occur for sufficiently energetic γ rays; another photon energy-loss process that is ordinarily forbidden is $\gamma \to \gamma + \pi^0$; and other potentially important processes include $\pi^0 \to N + \bar{N}$, which could become the dominant π^0 decay mode if it were energetically allowed. Typically, astrophysical observations involving observed quanta with energies E allow us to bound certain combinations of $k^{\mu\nu}_\pi$ components (with the $k^{\mu\nu}_\pi$ being a single Lorentz-violation coefficient tensor, common to all three pion fields) at the $\sim m_\pi^2/E^2$ level, or approximately 10^{-10}–10^{-13} in practice.

Placing bounds on Lorentz-violating effects that might be seen in pion propagation then immediately leads to similar bounds on Lorentz violation in pion–pion interactions. Because $\mathcal{L}'_\pi$ also contains a four-pion vertex of the form

$$\mathcal{L}'_{4\pi} = \frac{\beta^{(1)}}{12F^2}(c^{\mu\nu}_{u_L} + c^{\mu\nu}_{d_L} + c^{\mu\nu}_{u_R} + c^{\mu\nu}_{d_L})(\phi_a\phi_b\partial_\mu\phi_a\partial_\nu\phi_b - \phi_b\phi_b\partial_\mu\phi_a\partial_\nu\phi_a), \tag{10}$$

involving the same linear combination $k_\pi^{\mu\nu}$ as $\mathcal{L}'_{2\pi}$, observations of free mesons' dispersion relations provide direct information about their higher-order interaction vertices as well.

However, there are much stronger bounds on Lorentz violation in other hadronic sectors. Following a procedure that is analogous to the one used in the pion sector, it is possible to find the most general chirally-symmetric Lagrange density for the nucleon isodoublet.[2,3] Rather than a single $\mathcal{O}(1)$ low-energy constant, there are four in the baryonic χPT Lagrange density. Again, the chirally-invariant Lagrange density with the low-energy constants is equivalent to a standard SME Lagrange density—in this case, one containing $c^{\mu\nu}$ and $d^{\mu\nu}$ coefficients for the proton and neutron.

The nucleonic $c^{\mu\nu}$ and $d^{\mu\nu}$ coefficients are linear combinations of quark coefficients. For example,

$$c_p^{\mu\nu} = \frac{1}{2}\left[\alpha^{(1)}+\alpha^{(2)}\right](c_{u_L}^{\mu\nu}+c_{u_R}^{\mu\nu}) + \frac{1}{2}\left[-\alpha^{(1)}+\alpha^{(2)}\right](c_{d_L}^{\mu\nu}+c_{d_R}^{\mu\nu}), \quad (11)$$

expressing the proton $c^{\mu\nu}$ in terms of the same underlying quark coefficients as appeared in the pion sector, as well as two undetermined low-energy constants $\alpha^{(1)}$ and $\alpha^{(2)}$. Since most of the anisotropic c^{JK} coefficients for protons and neutrons have been bounded extremely tightly, this immediately leads to new bounds on the corresponding parameters for the pions. While the bounds on the k_π^{JK} based on analyses of high-energy astrophysical processes that would involve the pions directly were at the 10^{-10}–10^{-13} levels, the existence of bounds on baryonic Lorentz-violation coefficients leads to improvements to the 10^{-23}–10^{-24} levels.

Application of χPT has thus led to ten order of magnitude improvements in the bounds on certain SME parameters for mesons. With further analysis, including of Lorentz violation in the gluon sector,[4] and a better understanding of the low-energy constants $\alpha^{(i)}$ and $\beta^{(1)}$, further dramatic improvements can still be expected.

Acknowledgments

I wish to thank my collaborators, R. Kamand and M.R. Schindler.

References

1. D. Colladay and V.A. Kostelecký, Phys. Rev. D **58**, 116002 (1998).
2. R. Kamand, B. Altschul, and M.R. Schindler, Phys. Rev. D **95**, 056005 (2017).
3. R. Kamand, B. Altschul, and M.R. Schindler, Phys. Rev. D **97**, 095027 (2018).
4. J.P. Noordmans, Phys. Rev. D **95**, 075030 (2017).

Measuring the Fine-Structure Constant to Test Fundamental Laws of Physics

H. Müller, Z. Pagel, R.H. Parker, and W. Zhong

Department of Physics, University of California, Berkeley, CA 94720, USA

We have measured the fine-structure constant via the ratio h/m_{Cs} of the Planck constant and the mass of the cesium atom. In this proceedings, we will comment on prospects for further improvements.

1. The measurement

The fine-structure constant is ubiquitous in physics, and a comparison among different experiments provides a powerful test of the Standard Model of particle physics. We can use α to predict the anomaly $g_e - 2$ of the electrons magnetic moment. Comparison of theory and experiment of $g_e - 2$ is sensitive to a broad range of physics, both within the Standard Model and beyond, including Lorentz violation[1,2] and effects described by Standard Model effective field theory.[3]

We have measured $\alpha = 1/137.035\,999\,046(27)$ with an uncertainty of 200 parts per trillion (ppt), see Fig. 1.[4] A $2.5\,\sigma$ tension with the value obtained from $g_e - 2$ is a potential sign of new physics that mirrors the well-known $3.7\,\sigma$ tension observed in the muon $g_\mu - 2$. It motivates a deeper investigation at the frontier of precision measurements. The muon $g_\mu - 2$ is currently being re-measured by E989 at Fermilab, which expects to reduce the error more than threefold.[5] G. Gabrielse (Northwestern) is currently re-measuring $g_e - 2$ and expects an improvement by an order of magnitude to 20 ppt.[6]

The primary quantity being measured is h/m_{Cs}, the ratio of the Planck constant and the cesium mass. The fine-structure constant is determined using the relation[7]

$$\alpha^2 = \frac{2R}{c} \frac{m_{Cs}}{m_e} \frac{h}{m_{Cs}}. \qquad (1)$$

The experiment is based on an atom interferometer using matter waves of ^{133}Cs atoms, driven by the momentum $n\hbar k$ of n photons. The Rydberg

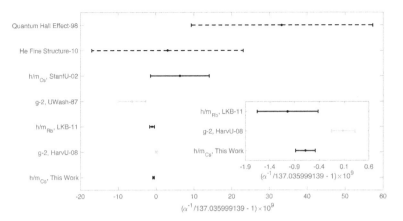

Fig. 1: Our recent result alongside previous measurements of α based on the quantum Hall effect, the He fine structure, h/m_{Cs}, h/m_{Rb},[7] and the electron $g_e - 2$[8]. Zero on the x axis is the CODATA 2014 recommended value.[9] Solid black bars indicate photon recoil experiments; gray ones electron $g_e - 2$ measurements.

constant[9] R contributes only an uncertainty of 3 ppt to α; the atom-to-electron mass ratio is expressed as $(m_{Cs}/\mathrm{amu})/(m_e/\mathrm{amu})$, where amu is the atomic mass unit. Currently, m_{Cs}/amu contributes 34 ppt.[10] K. Blaum (Heidelberg) is conducting a new measurement by mass spectroscopy with stored ions that will reduce this to 5 ppt.[11] The electron mass m_e/amu currently contributes 15 ppt,[12] which enables our future measurement to reach an accuracy of 20 ppt in α. We hope that future improvements will reduce this uncertainty threefold, so that we can determine the fine-structure constant with 10 ppt accuracy.

2. Planned improvements

Table 1 gives an overview of the error budget,[4] and of the improvements we intend to make. A few key improvements are discussed here.

Offset simultaneous conjugate interferometers (OSCIs).—While constant gravitational and other accelerations g are cancelled between two atom interferometers, acceleration gradients $\gamma_{zz} = \partial g_z/\partial z$ cause a relative error of $\approx \gamma T^2/12$ in the measurement of α. The effect of this first gradient will be cancelled by using a new interferometer geometry.[13]

Laser beam shape and quality.—While a plane electromagnetic wave at a frequency ω has a wave number of $k_0 = \omega/c$, a real laser beam has a spatially-varying, local effective wave number $k_{\mathrm{eff}} = k + 0 + \Delta k(x, y)$,

Table 1: Error budget in ppt.

Effect	Current	Future	Planned Improvement
laser frequency	30	2	
acceleration gradient	20.0	3.0	vertically offset AIs, modelling
Gouy phase	30.0	2.0	laser beam size and quality
wavefront curvature	30.0	2.0	laser beam size and quality
beam alignment	30.0	2.0	active stabilization
BO light shift	2.0	1.0	laser beam size and quality
density shift	3.0	1.0	dilute sample
index of refraction	30.0	1.0	dilute sample
speckle phase shift	40.0	2.0	Bloch oscillation launch, mode filtering
Sagnac effect	1.0	1.0	
mod. frequency	1.0	1.0	
thermal motion	80.0	1.3	laser beam size and quality
non-Gaussian waveform	30.0	0.5	laser beam size and quality
parasitic interferometers	30.0	0.8	larger signal
total systematic error	121.3	6.0	

where

$$\Delta k(x,y) = \frac{1}{2k_0^2}\frac{\nabla_\perp^2 E(x,y)}{E(x,y)},\qquad(2)$$

where $\nabla_\perp^2$ is the two-dimensional laplacian on a beam cross-section and $E(x,y)$ the local electric field amplitude.[14] This equation generalizes the well-known Gouy phase systematic for Gaussian beams. Since the photon momentum is proportional to k_{eff}, any deviations of the laser beam from constant intensity cause a systematic effect. We will improve the beam profile by using a thicker laser beam (25 mm instead of 3.21 mm radius), an oversized vacuum chamber tube to avoid stray reflections of the beam at the walls, and a spatial filter to improve further the beam profile.

Thermal motion.—Position dependence of the laser intensity causes a sensitivity to thermal motion of the atoms through intensity-dependent phase shifts in the laser–atom interaction. Intensity variations across the beam, coupled with thermal atom motion make these effects a function of the atom's time of flight. The residual systematic uncertainty was 80 ppt in the old experiment. Being generated by the intensity variations of the beam as a function of the beam radius, these effects should be suppressed to 1.3 ppt by using a thicker beam.

Statistical error.—The statistical error obtained in several days is equal to the accuracy that can be obtained in a measurement campaign of several months because the experiment has to be repeated many times at full sensitivity to characterize systematic errors. Atom detection with a CMOS camera will speed up the process, as the dependence of the result on the part of the cloud that is detected will be recorded with each experiment, without requiring repetitions. Similarly, OSCIs allow for substantial parallelization of systematic checks.

Acknowledgments

Stimulating discussions with A.V. Manohar are gratefully acknowledged.

References

1. *Data Tables for Lorentz and CPT Violation,* V.A. Kostelecký and N. Russell, 2019 edition, arXiv:0801.0287v12.
2. J.B. Araujo, R. Casana, and M.M. Ferreira, Jr., Phys. Rev. D **92**, 025049 (2015).
3. R. Alonso, E.E. Jenkins, A.V. Manohar, and M. Trott, J. High Energy Phys. **1404**, 159 (2014).
4. R.H. Parker, C. Yu, W. Zhong, B. Estey, and H. Müller, Science **360**, 191 (2018).
5. Muon g-2 Collaboration, J.L. Holzbauer *et al.*, PoS NuFact **2017**, 116 (2018).
6. G. Gabrielse, S.E. Fayer, T.G. Myers, and X. Fan, Atoms **7**, 45 (2019).
7. R. Bouchendira, P. Cladé, S. Guellati-Khélifa, F. Nez, and F. Biraben, Phys. Rev. Lett. **106**, 080801 (2011).
8. D. Hanneke, S. Fogwell, and G. Gabrielse, Phys. Rev. Lett. **100**, 120801 (2008).
9. P.J. Mohr, D.B. Newell, and B.N. Taylor, Rev. Mod. Phys. **88**, 035009 (2016).
10. G. Audi, F.G. Kondev, M. Wang, W.J. Huang, and S. Naimi, Chin. Phys. C **41**, 030001 (2017); E.G. Myers, Atoms **7**, 37 (2019).
11. K. Blaum, Phys. Rep. **425**, 1 (2006).
12. S. Sturm *et al.*, Nature **506**, 467 (2014).
13. W. Zhong, R.H. Parker, Z. Pagel, C. Yu, and H. Müller, arXiv:1901.03487.
14. S. Bade, L. Djadaojee, M. Andia, P. Cladé, and S. Guellati-Khélifa, Phys. Rev. Lett. **121**, 073603 (2018).

Maximal Tests in Minimal Gravity

Jay D. Tasson

Physics and Astronomy Department, Carleton College, Northfield, MN 55057, USA
LIGO-P1900177-v1

Recent tests have generated impressive reach in the gravity sector of the Standard-Model Extension. This contribution to the CPT'19 proceedings summarizes this progress and maps the structure of work in the gravity sector.

1. Lorentz violation in gravity

As demonstrated by the breadth of contributions to these proceedings and the ongoing growth of the *Data Tables for Lorentz and CPT Violation,*[1] the search for Lorentz violation as a signal of new physics, such as that originating at the Planck Scale,[2] is an active research area. The gravitational Standard-Model Extension (SME)[3–5] provides a field-theory-based framework for performing the search systematically. The structure of the SME can be thought of as a series expansion about known physics, with additional terms of increasing mass dimension constructed from conventional fields coupled to coefficients for Lorentz violation.[6] The leading terms, associated with operators of mass dimension $d = 3, 4$, are known as the minimal SME. In the gravity sector, phenomenology has been developed and tests have been performed based on a variety of complementary limits of the full SME. Relations among these efforts are summarized graphically in Fig. 1.

The framework for phenomenology in the gravity sector of the SME began in 2004 with Ref. 4, which developed the Lagrange density and associated theory to be used in searches for minimal Lorentz violation in gravity. Lorentz-violating effects in gravity can be understood as coming from the pure-gravity sector through Lorentz-violating modifications to the dynamics of the gravitational field,[7] or through gravitational couplings in Lorentz-violating terms in the other sectors of the theory.[8] In the latter case, Lorentz-violating effects are dependent on the species of matter contained in the test and source bodies, while in the former case they are not. References 7,8 address theory and phenomenology associated with minimal terms in pure gravity and matter-gravity couplings, respectively. A large

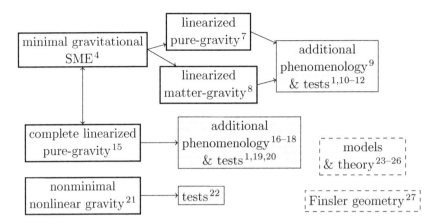

Fig. 1: Progress in SME gravity as of CPT'19. Light gray boxes show the various limits of the gravity sector that have been explored. Dark gray boxes show work that builds out the search in the respective limits. Dashed boxes show theoretical contributions.

amount of additional phenomenology[9] and experimental and observational searches[1] have been done based on these works, some of which are discussed in Sec. 2 and elsewhere in these proceedings.[10–12]

Some nonminimal gravity-sector terms were studied for short-range gravity experiments[13] and for gravitational Čerenkov radiation,[14] and the complete linearized theory of pure gravity was developed in Ref. 15, with initial applications to gravitational waves (GW). Since then, additional phenomenology[16–18] as well as experimental and observational work[1] has been done in nonminimal gravity. Examples of searches in nonminimal gravity are contained in these proceedings.[12,19,20] We note in passing the expected overlap between the linearized limit of the minimal work of Refs. 4,7 and the minimal limit of the complete linearized theory in Ref. 15. Study of the nonminimal gravity sector beyond the linearized limit has also begun.[21,22]

In addition to work aimed directly at seeking signals of Lorentz violation in experiments, several theory-oriented results deserve discussion in this context. While it is difficult to capture the volume of work done in this area in this short summary, examples discussed in these proceedings include exploration of specific Lorentz-violating models that generate nonzero SME coefficients[23,24] and the implications of geometric constraints on Lorentz violation.[25] The question of geometric constraints has also inspired consideration of Finsler geometry as a geometric framework for Lorentz violation.[27]

2. Maximal reach

Several recent and ongoing efforts have improved sensitivities to Lorentz violation in the minimal gravitational sector, or are expected to do so in the near future. A number of these are discussed elsewhere in these proceedings including improved sensitivities through matter–gravity couplings based on an analysis of data from the MICROSCOPE mission,[10] results from the analysis of solar-system data,[11] and tests based on interferometric gyroscopes.[12,28] Significant improvements in the laboratory were also achieved using gravimeters.[29] In this section, we summarize the recent effort providing the greatest reach, multimessenger astronomy.

On August 17, 2017, GWs and photons from the same astrophysical event were observed for the first time.[30] A gamma-ray burst arrived $(1.74 \pm 0.05)\,$s after the GWs from the coalescence of a pair of neutron stars. This observation, along with modeling suggesting up to a few seconds of lag between the coalescence and gamma-rays emission, led to a best-ever comparison of the speed of GWs and light. Such tests provide a particularly sensitive probe of $d = 4$ SME gravity coefficients due to the long propagation distance involved and because GW tests based on birefringence[15,17] and/or dispersion,[15,17,31] while powerful for $d > 4$, are insensitive to $d = 4$ coefficients. Using a maximum-reach analysis, in which the nine minimal $\overline{s}^{(4)}_{jk}$ gravity-sector coefficients are taken as nonzero one at time, the reach for all nine coefficients was improved over prior limits, most of which came from the analysis of Čerenkov radiation by cosmic rays.[14] The upper bound on the isotropic $\overline{s}^{(4)}_{00}$ coefficient was inaccessible to Čerenkov constraints, hence an improvement of ten orders of magnitude was achieved here, while improvements of up to a factor of 40 were achieved for the other coefficients. Future observations of multimessenger events offer several avenues of improvement. Events further away will improve the overall sensitivity, at least nine events distributed across the sky will enable the estimation of all nine $\overline{s}^{(4)}_{jk}$ coefficients together, and events at a variety of distances will disentangle speed differences from emission-time differences. The future is bright for seeking Lorentz violation with GWs.

Acknowledgments

J. Tasson is supported by NSF grant PHY1806990 to Carleton College.

References

1. *Data Tables for Lorentz and CPT Violation,* V.A. Kostelecký and N. Russell, 2019 edition, arXiv:0801.0287v12.

2. V.A. Kostelecký and S. Samuel, Phys. Rev. D **39**, 683 (1989).
3. D. Colladay and V.A. Kostelecký, Phys. Rev. D **58**, 116002 (1998).
4. V.A. Kostelecký, Phys. Rev. D **69**, 105009 (2004).
5. For a review, see J.D. Tasson, Rep. Prog. Phys. **77**, 062901 (2014); for a pedagogical introduction, see T.H. Bertschinger *et al.*, Symmetry **11**, 22 (2018).
6. For pedagogical discussion, see J.D. Tasson, JPS Conf. Proc. **18**, 011002 (2017).
7. Q.G. Bailey and V.A. Kostelecký, Phys. Rev. D **74**, 045001 (2006).
8. V.A. Kostelecký and J.D. Tasson, Phys. Rev. D **83**, 016013 (2011); Phys. Rev. Lett. **102**, 010402 (2009).
9. R.J. Jennings, J.D. Tasson, and S. Yang, Phys. Rev. D **92**, 125028 (2015); J.D. Tasson, Phys. Rev. D **86**, 124021 (2012); R. Tso and Q.G. Bailey, Phys. Rev. D **84**, 085025, (2011); Q.G. Bailey, Phys. Rev. D **80**, 044004 (2009).
10. G. Mo *et al.*, these proceedings; Q.G. Bailey *et al.*, in preparation.
11. C. Le Poncin-Lafitte *et al.*, these proceedings.
12. M.L. Trostel *et al.*, these proceedings; S. Moseley *et al.*, in preparation.
13. Q.G. Bailey, V.A. Kostelecký, and R. Xu, Phys. Rev. D **91**, 022006 (2015).
14. V.A. Kostelecký and J.D. Tasson, Phys. Lett. B **749**, 551 (2015).
15. V.A. Kostelecký and M. Mewes, Phys. Lett. B **757**, 510 (2016).
16. V.A. Kostelecký and M. Mewes, Phys. Lett. B **779**, 136 (2018).
17. M. Mewes, Phys. Rev. D **99**, 104062 (2019); these proceedings; K. O'Neal-Ault *et al.*, these proceedings.
18. Q.G. Bailey and D. Havert, Phys. Rev. D **96**, 064035 (2017).
19. C.-G. Shao *et al.*, these proceedings; C.-G. Shao *et al.*, Phys. Rev. Lett. **122**, 011102 (2019); Phys. Rev. Lett. **117**, 071102 (2016).
20. L. Shao, these proceedings; L. Shao and Q.G. Bailey, Phys. Rev. D **98**, 084049 (2018).
21. Q.G. Bailey, Phys. Rev. D **94**, 065029 (2016); these proceedings.
22. L. Shao and Q.G. Bailey, Phys. Rev. D **99**, 084017 (2019).
23. D. Colladay, these proceedings.
24. C.D. Lane, these proceedings; Q.G. Bailey and C.D. Lane, Symmetry **10**, 480 (2018).
25. R. Bluhm, these proceedings; Symmetry **9**, 230 (2017).
26. Y. Bonder *et al.*, these proceedings; N.A. Nilsson *et al.*, these proceedings.
27. B.R. Edwards, these proceedings; B.R. Edwards and V.A. Kostelecký, Phys. Lett. B **786**, 319 (2018).
28. For discussion of a relevant system, see A.D.V. Di Virgilio *et al.*, these proceedings.
29. N.A. Flowers, C. Goodge, and J.D. Tasson, Phys. Rev. Lett. **119**, 201101 (2017); C.-G. Shao *et al.*, Phys. Rev. D **97**, 024019 (2018).
30. LIGO, Virgo, Fermi GBM, and INTEGRAL Collaborations, B.P. Abbott *et al.*, Ap. J. Lett. **848**, L13 (2017).
31. A. Samajdar, these proceedings; LIGO and Virgo Collaborations, B.P. Abbott *et al.*, arXiv:1903.04467.

Test of CPT Symmetry with Positronium at the J-PET Detector

E. Czerwiński

Marian Smoluchowski Institute of Physics, Jagiellonian University
30-348 Kraków, Poland

On behalf of the J-PET Collaboration

The Jagiellonian Positron Emission Tomograph (J-PET) is the first tomograph built from plastic scintillators capable to test CPT symmetry in decays of positronium atoms. This test is performed via determination of the expectation value of operators that are odd under discrete symmetries. The J-PET detector is optimized for the registration of photons from electron–positron annihilations, and such operators may be constructed from the spin of ortho-positronium atoms and the momenta and polarization vectors of photons originating from its annihilation.

1. Positronium system

The bound state of an electron and a positron is an eigenstate of the charge-conjugation operator as well as a parity eigenstate. Such a bound state (positronium) has two possible ground states: $^1S_0$ para-positronium (p-Ps) and $^3S_1$ ortho-positronium (o-Ps) with 125 ps and 142 ns mean lifetime, respectively. It is the lightest purely leptonic object decaying into photons, and final-state interactions are expected to be at the level of 10^{-9} to 10^{-10}.[1,2] Therefore, it is a unique laboratory to study discrete symmetries.[3–6] In this paper, a possible CPT-symmetry test with the J-PET experiment[7] is discussed.

2. J-PET detector

A sodium source is placed in the center of the J-PET system. Emitted positrons are stopped in a porous material surrounding the source[8] creating o-Ps atoms, which may annihilate into 3γ. A dedicated trilateration method[9] is used for the annihilation-place and -time determination (taking into account the coplanarity of the momentum vectors of the produced gamma quanta), while the reconstructed directions of the annihilation photons are used for calculating their energy.[10] The reconstructed annihilation point and the known source position allow for the determination of the e^+ velocity direction, which also yields the spin direction of the positron. Since the polarization of positrons is to a large extent preserved during the thermalization process,[11,12] the spin of o-Ps is determined as well.[13]

J-PET is a cylindrical multipurpose detector[14,15] composed of three layers (with radii of 425, 467.5, and 575 mm) of plastic scintillator strips. The position of the photon interaction along the scintillator strip is derived from the time difference of the signals from both photomultipliers attached to this strip,[16] whereas the time of interaction is calculated from the sum of times of these signals.[17,18] In total, up to eight measurements of each signal are performed in the time domain at four different amplitude thresholds.[19,20] This allows for Time-Over-Threshold measurements (equivalent to the determination of the energy deposition by the annihilation photon) and a precise extraction of the signal start time. The data stream is processed in a triggerless manner and reconstructed offline with our developed analysis framework.[21]

At J-PET, a polymer scintillator is used. Therefore, the photon deposits its energy via Compton scattering. The registration of primary $(\vec{k_i})$ and scattered $(\vec{k_i'})$ gamma pairs allows the estimation of the linear polarization $\vec{\epsilon_i} = \vec{k_i} \times \vec{k_i'}$,[22,23] which is an extraordinary feature of the J-PET system.[7,24] Details about J-PET characteristics are presented in Refs. 25–27.

3. CPT studies with J-PET

Operators with expectation values accessible at J-PET for discrete-symmetry studies with the o-Ps $\rightarrow 3\gamma$ decay are presented in Table 1. The most precise determination of the amplitude for a CPT-violating asymmetry in the decays of positronium was achieved by the Gammasphere detector:[28]

$$C_{CPT} = 0.0026 \pm 0.0031, \tag{1}$$

for the expectation value $\langle \vec{S} \cdot (\vec{k_1} \times \vec{k_2}) \rangle$. Therefore, there is at least a six orders of magnitude level for the possible discovery of CPT-symmetry violation taking into account the expected level of final-state interactions.[1,2] The expected accuracy for measurements with the J-PET system is $\mathcal{O}(10^{-5})$ for CPT-odd expectation values.[7] A Standard-Model Extension (SME)[29] analysis can also be performed since the experimental data are collected during long measurement campaigns in the continuous 24 h/day trigger-less mode. Detector performance and stability were confirmed in a continuous data-

Table 1: Operators for the o-Ps $\rightarrow 3\gamma$ decay and their parity under discrete transformations. The momentum of the ith photon is denoted by $\vec{k}_i$ with $k_1 \geq k_2$, $\vec{S}$ is the spin of the o-Ps, and the polarization of ith photon is $\vec{\epsilon}_i$.[7] The notation for the operator behavior under each of the listed transformations is + and − for even and odd parity, respectively. Operators involving the polarization $\vec{\epsilon}_i$ are exclusively available at the J-PET system.

Operator	C	P	T	CP	CPT
$\vec{S} \cdot \vec{k}_1$	+	−	+	−	−
$\vec{S} \cdot (\vec{k}_1 \times \vec{k}_2)$	+	+	−	+	−
$(\vec{S} \cdot \vec{k}_1)[\vec{S} \cdot (\vec{k}_1 \times \vec{k}_2)]$	+	−	−	−	+
$\vec{k}_1 \cdot \vec{\epsilon}_2$	+	−	−	−	+
$\vec{S} \cdot \vec{\epsilon}_1$	+	+	−	+	−
$\vec{S} \cdot (\vec{k}_2 \times \vec{\epsilon}_1)$	+	−	+	−	−

collection period reaching four months. Thus far, J-PET completed seven data-taking campaigns with the eighth one currently ongoing.

A fourth detection layer will be installed inside the J-PET system soon. This will increase the geometrical acceptance providing full coverage in azimuthal angle.

Acknowledgments

This work was supported by the Polish National Center for Research and Development under grant No. INNOTECH-K1/IN1/64/159174/NCBR/12, by the Foundation for Polish Science under the MPD and TEAM/2017-4/39 programs, by the National Science Center of Poland under grant Nos. 2016/21/B/ST2/01222 and 2017/25/N/NZ1/00861, and by the Ministry for Science and Higher Education under grant Nos. 6673/IA/SP/2016, 7150/E-338/SPUB/2017/1, 7150/E-338/M/2017, and 7150/E-338/M/2018.

References

1. W. Bernreuther, U. Low, J.P. Ma, and O. Nachtmann, Z. Phys. C **41** 143 (1988).
2. B.K. Arbic *et al.*, Phys. Rev. A **37**, 3189 (1988).
3. S.D. Bass, accepted for publication in Acta Phys. Pol. B, arXiv:1902.01355.
4. D.B. Cassidy, Eur. Phys. J. D **72**, 53 (2018).
5. S.G. Karshenboim, Phys. Rept. **422**, 1 (2005).
6. S.N. Gninenko, N.V. Krasnikov, and A. Rubbia, Mod. Phys. Lett. A **17**, 1713 (2002).

7. P. Moskal *et al.*, Acta Phys. Pol. B **47**, 509 (2016).
8. B. Jasińska *et al.*, Acta Phys. Polon. B **47**, 453 (2016).
9. A. Gajos *et al.*, Nucl. Instrum. Meth. A **819**, 54 (2016).
10. D. Kamińska *et al.*, Eur. Phys. J. C **76**, 445 (2016).
11. P.W. Zitzewitz *et al.*, Phys. Rev. Lett. **43**, 1281 (1979).
12. J. Van House and P.W. Zitzewitz, Phys. Rev. A **29**, 96 (1984).
13. M. Mohammed *et al.*, Acta Phys. Pol. A **132**, 1486 (2017).
14. P. Moskal *et al.*, Phys. Med. Biol. **64**, 055017 (2019).
15. P. Moskal *et al.*, Phys. Med. Biol. **61**, 2025 (2016).
16. P. Moskal *et al.*, Nucl. Instrum. Meth. A **764**, 317 (2014).
17. L. Raczyński *et al.*, Phys. Med. Biol. **62**, 5076 (2017).
18. L. Raczyński *et al.*, Nucl. Instrum. Meth. A **786**, 105 (2015).
19. M. Pałka *et al.*, JINST **12**, P08001 (2017).
20. G. Korcyl *et al.*, IEEE Trans. Med. Imag. **37**, 2526 (2018).
21. W. Krzemień *et al.*, Acta Phys. Pol. B **47**, 561 (2016).
22. O. Klein and T. Nishina, Z. Phys. **52**, 853 (1929).
23. R.D. Evans in S. Flügge, ed., *Corpuscles and Radiation in Matter II*, Encyclopedia of Physics **6**, Springer Verlag, Berlin, 1958.
24. P. Moskal *et al.*, Eur. Phys. J. C **78**, 970 (2018).
25. E. Czerwiński *et al.*, Acta Phys. Pol. B **48**, 1961 (2017).
26. P. Kowalski *et al.*, Phys. Med. Biol. **63**, 165008 (2018).
27. S. Niedźwiecki *et al.*, Acta Phys. Polon. B **48**, 1567 (2017).
28. P.A. Vetter and S.J. Freedman, Phys. Rev. Lett. **91**, 263401 (2003).
29. *Data Tables for Lorentz and CPT Violation*, V.A. Kostelecký and N. Russell, 2019 edition, arXiv:0801.0287v12.

Tests of Lorentz Invariance at the Sudbury Neutrino Observatory

K.R. Labe

Department of Physics, Cornell University, Ithaca, NY 14853, USA

On behalf of the SNO Collaboration

Seven years of solar neutrino data collected at the Sudbury Neutrino Observatory are analyzed for possible signals of Lorentz violation, which would present as either seasonal variations in the electron-neutrino survival probability, or as a shape change to the oscillated energy spectrum. No evidence for Lorentz violation was observed in the data, so constraints on such effects were established in the framework of the Standard-Model Extension, including 38 limits on previously unconstrained operators and improved limits on 16 additional operators. As a result of this work, experimental limits on all minimal, Dirac-type Lorentz-violating operators in the neutrino sector are now available.

1. Introduction

The great power of searching for Lorentz violation in solar neutrinos derives from the existence of the matter or MSW[1] effect in the Sun, which causes solar neutrinos to be produced in a state that is a mixture of the vacuum flavor eigenstates. As a result, measurements on solar neutrinos can provide information about Lorentz violation in all flavor components. If Lorentz violations were present in the neutrino sector, one would expect a directional dependence to neutrino propagation, which would lead to seasonal variations in the solar electron-neutrino survival probability.

The Sudbury Neutrino Observatory (SNO)[2] was able to make a precise measurement of the solar electron-neutrino survival probability through its distinct flavor-tagging[3] and flavor-neutral detection channels.[4] These data were used to search for time-of-year variations in the survival probability.[5]

2. SNO detector

SNO was a heavy-water Cherenkov detector located at a depth of 2100 m (5890 m.w.e.) in Vale's Creighton mine near Sudbury, Ontario. The detector consisted of a number of nested volumes. At the center were 1000 metric tons of ^{2}H$_2$O held in a 12-m diameter spherical acrylic vessel. Outside this was a 17.8-m diameter geodesic support structure, which held 9456 20-cm photomultiplier tubes. The entire detector was suspended in a barrel-shaped cavity filled with ultra-pure water to act as shielding against background radiation.

SNO was sensitive to three solar neutrino interaction channels:

$$
\begin{aligned}
\nu_e + d &\rightarrow p + p + e^- - 1.44\,\text{MeV} &&\text{(CC)}, \\
\nu + d &\rightarrow p + n + \nu - 2.22\,\text{MeV} &&\text{(NC)}, \\
\nu + e^- &\rightarrow \nu + e^- &&\text{(ES)}.
\end{aligned}
$$

The neutral-current (NC) interaction couples to neutrinos of all flavors equally and allowed an inclusive measurement of the solar neutrino flux. The charged-current (CC) and elastic-scattering (ES) interactions couple exclusively (CC) or preferentially (ES) to the electron-flavor neutrino, which allowed the solar electron-neutrino survival probability to be measured.

3. Lorentz violation for solar neutrinos

The theory of Lorentz violations in the neutrino sector is well established.[6] In this analysis, it is assumed that only the a- and c-type coefficients, which control mixing among neutrinos, are present as a small perturbation to the Standard Model mass terms, while the g and H coefficients, which would induce mixing between neutrinos and antineutrinos, are negligibly small. In this limit, the impact on the solar electron-neutrino survival probability can be summarized[5] at first order in perturbation theory as

$$
\delta P_{ee}^{(1)} = \mathfrak{Re} \sum_{jm} Y_{jm}(\hat{p}) \left(E \left(a_{\text{SNO}}^{(3)} \right)_{jm} - E^2 \left(c_{\text{SNO}}^{(4)} \right)_{jm} \right), \tag{1}
$$

where Y_{jm} are the spherical-harmonic functions, $\hat{p}$ is the unit vector pointing from the Sun to the Earth, E is the neutrino energy, and a_{SNO} and c_{SNO} are linear combinations[5] of the different flavor components representing the sensitivity of SNO.

4. Competing effects

Two known effects that give rise to seasonal variations in the solar electron-neutrino flux were identified: changes in flux due to the eccentricity of the Earth's orbit, and seasonal perturbations to the survival probability caused by electron-neutrino regeneration as a result of the matter effect in the

Earth.[7] The effect from eccentricity was removed by exactly calculating its impact in the analysis. The Earth matter effect is significantly more complicated to evaluate, since a detailed model of the electron number density inside the Earth is required. Instead of removing this effect, a systematic uncertainty was applied.

5. Analysis

For the analysis,[5] strong data-selection cuts were applied, particularly in energy. Only events reconstructing in the range of 7–20 MeV were included in the analysis. These stringent cuts were developed to reduce the number of background events to a very low level, estimated to be about 2%. This was necessary because strong constraints on the stability of the background rates as a function of time could not be developed. Because the external water was recirculated during the course of the experiment, background rates did fluctuate, and we could establish stability only at a level of 50%.

The analysis, conducted blinded, was performed as a binned likelihood fit in energy, volume-weighted radius, solar angle, and isotropy. The fits had three free parameters: the solar neutrino flux and mixing angle, and one Lorentz violation parameter. The fit was repeated eight times for the eight independent Lorentz-violation signal types.

Many possible systematic effects were considered including uncertainties in the shape and normalization of the PDFs used in the fit, time variations in background-event rates, and uncertainties in the neutrino mixing model. Because the measurement was ultimately statistically limited, the systematic error was estimated using a shift-and-refit strategy. The dominant systematic came from the uncertainty in the stability of the background data, which ultimately contributed about 0.3 times the statistical uncertainty.

6. Results

No evidence for Lorentz violation was observed in the data. As a result, limits were established on the sizes of the different effects. Limits on the individual flavor components could be calculated in a "maximal reach" framework by combining the limits on the different signal types with information about the flavor sensitivity of the experiment. Care must be taken since these two pieces of information are correlated.[5] Limits at the 95% CL on the a terms range from 10^{-20} to 10^{-21} GeV, while the limits on c terms range from 10^{-17} to 10^{-19}. The results for each individual flavor component can be found in Ref. 5.

24

Acknowledgments

This research was supported by Canada: Natural Sciences and Engineering Research Council, Industry Canada, National Research Council, Northern Ontario Heritage Fund, Atomic Energy of Canada, Ltd., Ontario Power Generation, High Performance Computing Virtual Laboratory, Canada Foundation for Innovation, Canada Research Chairs program; US: Department of Energy Office of Nuclear Physics, National Energy Research Scientific Computing Center, Alfred P. Sloan Foundation, National Science Foundation, the Queen's Breakthrough Fund, Department of Energy National Nuclear Security Administration through the Nuclear Science and Security Consortium; United Kingdom: Science and Technology Facilities Council (formerly Particle Physics and Astronomy Research Council); Portugal: Fundação para a Ciência e a Tecnologia. We thank the SNO technical staff for their strong contributions. We thank INCO (now Vale, Ltd.) for hosting this project in their Creighton mine.

References

1. L. Wolfenstein, Phys. Rev. D **17**, 2369 (1978); S.P. Mikheyev and A.Y. Smirnov, Nuovo Cimento C **9**, 17 (1986).
2. SNO Collaboration, J. Boger *et al.*, Nucl. Instrum. Meth. A **449**, 172 (2000).
3. SNO Collaboration, Q.R. Ahmad *et al.*, Phys. Rev. Lett. **87**, 071301 (2001).
4. SNO Collaboration, Q.R. Ahmad *et al.*, Phys. Rev. Lett. **89**, 011301 (2002).
5. SNO Collaboration B. Aharmim, Phys. Rev. D **98**, 112013 (2018).
6. V.A. Kostelecký and M. Mewes, Phys. Rev. D **85**, 096005 (2012).
7. S.S. Aleshin, O.G. Kharlanov, and A.E. Lobanov, Phys. Rev. D **87**, 045025 (2013); O.G. Kharlanov, arXiv:1509.08073.

Symmetries in the SME Gravity Sector: A Study in the First-Order Formalism

Y. Bonder[*] and C. Corral[*,†]

Instituto de Ciencias Nucleares, Universidad Nacional Autónoma de México, Circuito Exterior s/n Ciudad Universitaria, Ciudad de México, 04510, Mexico

†*Universidad Andrés Bello, Departamento de Ciencias Físicas, Facultad de Ciencias Exactas, Sazié 2212, Santiago, 8370136, Chile*

A method to find the symmetries of a theory in the first-order formalism of gravity is presented. This method is applied to the minimal gravity sector of the Standard-Model Extension. It is argued that no inconsistencies arise when Lorentz violation is explicit and the relation between Lorentz violation and invariance under (active) diffeomorphisms is clearly exposed.

The Standard-Model Extension[1] (SME) is a framework to parameterize all possible violations of local Lorentz invariance. It has a gravitational sector[2] where the fields describing the spacetime geometry are coupled to the SME coefficients. Here, only explicit Lorentz violation is considered where the SME coefficients are nondynamical.

On the other hand, the first-order formalism of gravity[3] has the vierbein and an independent Lorentz connection as the dynamical geometrical variables. The mathematical framework is that of differential p-forms, i.e., totally antisymmetric tensors of rank $(0,p)$, $p \leq 4$. In fact, the vierbein e^a is an $\mathfrak{so}(1,3)$ valued 1-form (Latin indices are Lorentz indices; spacetime indices are omitted) and the Lorentz connection is a 1-form $\omega^{ab} = -\omega^{ba}$.

The basic operations of this framework are the wedge product $\wedge$, the inner product with respect to the vector field ξ, i_ξ, and the exterior derivative d. Basically, $\wedge$ is a tensor product whose result is antisymmetrized, i_ξ saturates the p-form with ξ, thus reducing the rank of the form by one, and d is an antisymmetrized derivative using any torsionless derivative operator that raises the rank of the form by one. In addition, the Minkowski metric $\eta_{ab} = \text{diag}(-1,1,1,1)$ and its inverse η^{ab} are used to lower and raise Lorentz indices and the summation convention over repeated indices is assumed. The conventions that are used are those of Ref. 4.

The central equations of this formalism are Cartan's structure equations for the curvature and torsion 2-forms, R^{ab} and T^a:

$$R^{ab} = \mathrm{d}\omega^{ab} + \omega^a{}_c \wedge \omega^{cb}, \quad T^a = \mathrm{d}e^a + \omega^a{}_b \wedge e^b. \tag{1}$$

Moreover, the Lorentz connection can be used to define a covariant exterior derivative D. This is done by taking the conventional exterior derivative and adding (subtracting) a term for each superscripted (subscripted) Lorentz index. By definition, D is Lorentz covariant. Using that $\mathrm{d}^2 = 0$, it is easy to show that $\mathrm{D}R^{ab} = 0$ and $\mathrm{D}T^a = R^a{}_b \wedge e^b$, which are the Bianchi identities.

The main advantages of the first-order formalism are its suitability for integration and its efficiency regarding action variations. It also considers a more general connection that is *a priori* torsion full; to recover the results of the metric formalism one needs to set torsion to zero consistently. Moreover, the Lie derivative along a vector field ξ, when acting on a p-form θ, simply becomes $\mathcal{L}_\xi \theta = i_\xi \mathrm{d}\theta + \mathrm{d}i_\xi \theta$; this is known as Cartan's formula.[4]

At this point, attention is focused on applications of the first-order formalism to the SME. The minimal part of the gravitational sector of the SME (mgSME) has Lorentz-violating terms modifying the action of General Relativity (GR) with no additional derivatives. In vacuum and in the first-order formalism the corresponding action is[5]

$$S_{\mathrm{mgSME}}[\omega^{ab}, e^a] = \tfrac{1}{2\kappa} \int \left(\epsilon_{abcd} + k_{abcd} \right) R^{ab} \wedge e^c \wedge e^d, \tag{2}$$

where κ is the gravitational coupling constant, ϵ_{abcd} is the totally antisymmetric Lorentz tensor, and k_{abcd} is a nondynamical 0-form that plays the role of the SME coefficients. The term proportional to ϵ_{abcd} corresponds to the Einstein–Hilbert action in the presence of torsion (or the Einstein–Cartan action) without a cosmological constant. Interestingly, due to the presence of torsion, there are more coefficients than in the conventional mgSME. In particular, k_{abcd} is such that $k_{abcd} = k_{[ab][cd]}$ but $k_{abcd} \neq k_{cdab}$, reflecting the fact that the Ricci tensor is not symmetric (squared brackets denote antisymmetrization with a factor $1/2$).

An arbitrary variation of compact support yields

$$\delta S_{\mathrm{mgSME}} = \int \left(\delta\omega^{ab} \wedge \mathcal{E}_{ab} + \delta e^a \wedge \mathcal{E}_a \right), \tag{3}$$

$$\mathcal{E}_{ab} = \tfrac{1}{\kappa} \left(\epsilon_{abcd} + k_{abcd} \right) T^c \wedge e^d + \tfrac{1}{2\kappa} \mathrm{D}k_{abcd} \wedge e^c \wedge e^d, \tag{4}$$

$$\mathcal{E}_a = \tfrac{1}{\kappa} \left(\epsilon_{abcd} + k_{abcd} \right) R^{bc} \wedge e^d. \tag{5}$$

Clearly, $\mathcal{E}_{ab} = 0 = \mathcal{E}_a$ are the equations of motion (EOM). Thus, $\mathcal{E}_{ab}$ and $\mathcal{E}_a$ are called EOM throughout the text. However, the symmetries must be studied off shell and the EOM are not assumed to vanish. The covariant exterior derivatives of the EOM can be cast into the form

$$D\mathcal{E}_{ab} = e_{[a} \wedge \mathcal{E}_{b]} + \frac{1}{\kappa}\left(- 2k_{[a|cde|}R^{cd} \wedge e_{b]} \wedge e^e \right.$$

$$\left. + k_{abcd}R^c{}_e \wedge e^e \wedge e^d - k_{[a|ecd|}R^e{}_{b]} \wedge e^c \wedge e^d \right), \tag{6}$$

$$D\mathcal{E}_a = i_a T^b \wedge \mathcal{E}_b + i_a R^{bc} \wedge \mathcal{E}_{bc} - \frac{1}{\kappa} i_a Dk_{lbcd} \wedge R^{bc} \wedge e^l \wedge e^d, \tag{7}$$

where i_a is such that $i_\xi = \xi^a i_a$. Importantly, all term that cannot be written as some tensor contracted with the EOM are called symmetry-breaking terms (SBT), and, in this case, the SBT are those terms with a k_{abcd}.

The symmetries and the resulting conditions can be read off from Eqs. (6) and (7) using the following method (see Ref. 6): Step 1, multiply these equations by the 'gauge parameters,' namely by Lorentz-valued 0-forms of compact support $\lambda^{ab} = -\lambda^{ba}$ and ξ^a, respectively. Step 2, integrate over spacetime and use the Leibniz rule to convert $\lambda^{ab}D\mathcal{E}_{ab}$ and $\xi^a D\mathcal{E}_a$ to $D\lambda^{ab} \wedge \mathcal{E}_{ab}$ and $D\xi^a \wedge \mathcal{E}_a$. The resulting equations are

$$0 = \int \left[D\lambda^{ab} \wedge \mathcal{E}_{ab} - \lambda^{ab} e_b \wedge \mathcal{E}_a + \frac{\lambda^{ab}}{\kappa}\left(- 2k_{acde}R^{cd} \wedge e_b \wedge e^e \right. \right.$$

$$\left. \left. + k_{abcd}R^c{}_e \wedge e^e \wedge e^d - k_{aecd}R^e{}_b \wedge e^c \wedge e^d \right) \right], \tag{8}$$

$$0 = \int \left[i_\xi R^{ab} \wedge \mathcal{E}_{ab} + (D\xi^a + i_\xi T^a) \wedge \mathcal{E}_a - \frac{1}{\kappa} i_\xi Dk_{abcd} \wedge R^{bc} \wedge e^a \wedge e^d \right]. \tag{9}$$

Step 3, check if there exists a nontrivial gauge parameter $\tilde{\lambda}^{ab}$ ($\tilde{\xi}^a$) such that the SBT in Eq. (8) (Eq. (9)) vanish. If this occurs, the equation takes the form of the action variation (3) and, by comparison, it is possible to read off the field transformations: $\delta\omega^{ab} = D\tilde{\lambda}^{ab}$ and $\delta e^a = -\tilde{\lambda}^{ab} e_b$ ($\delta\omega^{ab} = i_{\tilde{\xi}} R^{ab}$ and $\delta e^a = D\tilde{\xi}^a + i_{\tilde{\xi}} T^a$). Conversely, if there are no gauge parameters such that the SBT vanish, then the theory has no symmetries.

Note that the transformation laws for the vierbein and the Lorentz connection obtained from Eq. (8) coincide with the well-known Lorentz transformations. Since, for an arbitrary k_{abcd}, there is no nontrivial $\tilde{\lambda}^{ab}$ such that the SBT vanish, it is possible to conclude that this symmetry is completely broken in the mgSME. On the other hand, the transformations arising from Eq. (9) are not diffeomorphisms (Diff), as one could naively

28

expect, but are covariant Diff in which, in Cartan's formula, d is replaced by D. This suggests that the fundamental symmetries of the theories under consideration, including GR, are not the conventional Diff but their covariant version. Again, for an arbitrary k_{abcd} there is no $\tilde{\xi}^a$ for which the SBT vanish. Therefore, the mgSME breaks this symmetry. Notice, however, that one can have invariance under the covariant Diff (e.g., D$k_{abcd} = 0$), but the theory can still break the conventional Diff invariance since the Diff are a combination of a covariant Diff and a Lorentz transformation. Another interesting example is the unimodular version of Einstein–Cartan theory,[6,7] which is Lorentz invariant but is not invariant under all Diff.

In conclusion, a rigorous method to find the symmetries in a particular theory has been presented. With this method, it can be shown[5] that no inconsistencies arise between the Bianchi identities and a nondynamical k_{abcd}. At most, the conservation laws impose restrictions on k_{abcd}, which goes against the typical SME working hypothesis. Another lesson from this analysis is that the interplay of Lorentz and Diff violation is richer than is usually considered. Other possible applications of the first-order formalism in the context of explicit Lorentz violation can include the construction of the hamiltonian by following the method of Ref. 8 and the construction of nonminimal terms using the fact that 4-forms are the natural objects in the action, where the Hodge dual will certainly play a role.

Acknowledgments

This work was supported by UNAM-DGAPA-PAPIIT Grant IA101818.

References

1. D. Colladay and V.A. Kostecký, Phys. Rev. D **55**, 6760 (1997); Phys. Rev. D **58**, 116002 (1998).
2. V.A. Kostelecký, Phys. Rev. D **69**, 105009 (2004); Y. Bonder, Phys. Rev. D **91**, 125002 (2015); Y. Bonder and G. León, Phys. Rev. D **96**, 044036 (2017).
3. M. Blagojević and F.W. Hehl, eds., *Gauge Theories of Gravitation*, World Scientific Publishing, 2013.
4. M. Nakahara, *Geometry, Topology and Physics*, Taylor & Francis, 2016.
5. Y. Bonder and C. Corral, Symmetry **10**, 433 (2018).
6. Y. Bonder and C. Corral, Phys. Rev. D **97**, 084001 (2018).
7. C. Corral and Y. Bonder, Class. Quantum Grav. **36**, 045002 (2019).
8. J.M. Nester, Mod. Phys. Let. A **6**, 2655 (1991); C.M. Chen, J.M. Nester, and R.S. Tung, Int. J. Mod. Phys. D **24**, 1530026 (2015).

Limits on Lorentz-Invariance Violation from HAWC

James T. Linnemann

Department of Physics and Astronomy, Michigan State University,
East Lansing, Michigan 48824, USA

On behalf of the HAWC Collaboration

The High Altitude Water Cherenkov observatory has observed the highest energy photons ever measured. This can place strongly constraining limits on Lorentz violation via the photon decay channel, which are directly comparable to superluminal limits from time-of-arrival differences of photons from different energies, but a factor of ten stronger.

1. HAWC

The High Altitude Water Cherenkov Observatory (HAWC) is located at the Sierra Negra volcano in the state of Puebla, Mexico, at an altitude of 4100 m. It is a wide-field observatory consisting of 300 tanks of 200,000 L of water, each containing four photomultiplier tubes. HAWC measures TeV extensive air showers and classifies the showers as induced by photons or hadrons using shower characteristics. HAWC has operated at 95% duty cycle since 2015. Recently, we have inaugurated a neural-net-based energy-estimation technique with a resolution of 0.1 or better in $\log_{10}(E/\text{TeV})$, as shown in the left panel of Fig. 1. This improved energy resolution allows us to measure energy spectra of astrophysical photon sources, such as the Crab Nebula and other TeV emitters. HAWC is particularly sensitive to sources with spatially extended emission because its large instantaneous field of view (2 sr) allows using extended regions of the sky in its background estimation. More detail on HAWC, the neural-net-based energy measurement technique, and its application to the Crab Nebula energy spectrum can be found in Ref. 1.

2. High-energy sources

Seven sources are used in this analysis; they are listed in Table 1. All but the Crab Nebula are within $2°$ of the galactic plane. These sources are chosen because their spectra include significant high-energy emission above 56 TeV in reconstructed energy. The Crab Nebula was modeled as a point source with a log-parabola spectrum. The others were modeled with exponentially cut-off power-law spectra. These sources are spatially extended and modeled by a Gaussian morphology with radius fixed at best-fit values found during the construction of HAWC's high-energy catalog.[2] The right panel of Fig. 1 shows the energy spectrum of the most energetic of these sources.

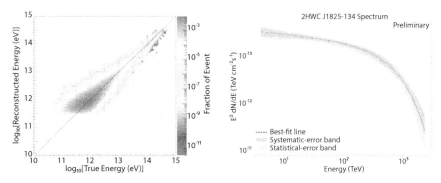

Fig. 1: Left: Reconstructed vs. true energy. Right: Spectrum of 2HWC J1825-134 with curvature evident above 100 TeV.

3. Photon decay effects on source energy spectra

The Lorentz-violation (LV) analysis is performed in the context of a generic modified LV dispersion relationship for the photon:[3]

$$E^2 - (1 + \alpha_n p^n)p^2 = 0. \tag{1}$$

Here, α parameterizes a small violation of $E^2 - p^2 = 0$. Its translation to a quantum-gravity scale is $EQG_n = 1/\sqrt[n]{\alpha_n}$, in which the correction to the p^2 term would be written as $(p/EQG_n)^n$. We only consider $n = 1, 2$ in this work. When α_n is positive, $E^2 - p^2 > 0$, and photon decay into an e^+e^- pair is allowed for photon momenta $p^{n+2} > (2m_e)^2/\alpha_n$. QED calculations of the decay rate find that the decay occurs with essentially with 100% probability and very near the source.

4. Setting limits on photon decay

The resulting photon-decay phenomenology is an abrupt termination in the intrinsic source spectrum at some energy E_{LV}. However, this abrupt cutoff

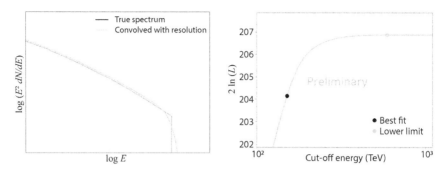

Fig. 2: Left: Spectrum used in the LV fit for the Crab Nebula. The effect of the reconstructed energy resolution is shown. Right: Likelihood curve vs. cutoff energy for the Crab Nebula with 95% c.l.; J1825 is similar.

will be softened in the measured energy due to detector energy resolution. Thus, to bound LV, the next step is to see whether the observed spectra are well described by the spectral shape shown in left panel of Fig. 2. We find the logarithmic-profile likelihood ratio between the best-fit spectrum with LV cutoff $\hat{E}$ and the best fit without LV cutoff (i.e., the cutoff set to ∞)

$$D = 2\ln\{L(\hat{E})/L(\infty)\}. \qquad (2)$$

We calculate a p value of observing D or higher assuming the null hypothesis of no LV. The result is shown for each source in Table 1. None of the sources favors a cutoff, since the smallest p value is only 0.294, which is well below a 2σ-equivalent deviation from the smooth spectrum.

Thus, we proceed to setting a lower limit on E_{LV} at 95% c.l. This is done by varying E_{LV} below its best-fit value $\hat{E}$ until the logarithmic likelihood $2\ln L(E)/L(\hat{E})$ decreases by 2.71 units below the best-fit value $2\ln L(\hat{E})$.[4] The right panel of Fig. 2 shows the likelihood as a function of E_{LV} for the Crab Nebula along with the best-fit value and the 95% c.l. value. The results are shown in Table 1. Clearly, if a larger confidence-level content than 95% is required, the resulting limit will decrease significantly. The E_{LV} column can be directly interpreted as a 95% c.l. lower limit on observed photons from each source which can be used to constrain astrophysical-source acceleration models. The method makes no assumptions about the spectrum shape *above* the limit, only that the chosen energy-spectrum model represents well the spectrum below the cutoff.

The results in Table 1 represent the continuation of a Ph.D. thesis,[5] but are based on improved HAWC MC simulations.[1] A future publication will take into account systematic uncertainties. For comparison, results derived from the previous-best Crab Nebula spectrum by HEGRA and the limits placed using Fermi GRB data are also shown. The HAWC data set a higher limit than previous Crab Nebula data because the HAWC spectra extend to higher energy. The EQG values scale with observed photon energy as E^{4-n}, so the new limits are improved by up to two orders of magnitude. The EQG values for $n=1$ easily exceed the Planck energy 0.12×10^{30} eV. The Fermi limits are based on energy-dependent photon propagation time.[6]

Table 1: Preliminary HAWC sources and 95% c.l. LV limits.

Source	E_{LV} TeV	EQG_1 10^{30} eV	EQG_2 10^{22} eV	α_1 10^{-31}/eV	α_2 10^{-45}/eV^2	p Value
2HWC J1825-134	253	15.5	6.26	0.64	0.26	1.000
2HWC J1908+063	213	9.3	4.44	1.08	0.51	0.990
Crab (HAWC)	152	3.4	2.26	2.97	1.96	1.000
2HWC J2031+415	144	2.9	2.02	3.5	2.43	0.714
2HWC J2019+367	121	1.7	1.43	5.6	4.87	0.8282
J1839-057	79	0.47	0.61	21.1	26.8	0.357
2HWC J1844-032	77	0.44	0.58	22.9	29.7	0.294
Crab (HEGRA)[3]	56	0.15	0.28			
Fermi GRB $v > c$[6]	n/a	0.080	0.009			
Fermi GRB $v > c$[6]	n/a	13.2	0.009			
Fermi GRB $v < c$[6]	n/a	0.91	0.013			

Acknowledgments

The author thanks Alan Kostelecký for his generous hospitality.

References

1. HAWC Collaboration, A.U. Abeysekara *et al.*, arXiv:1905.12518.
2. HAWC Collaboration, *The HAWC Observatory High-Energy Astrophysical Source Catalog*, to appear; K. Malone, Pennsylvania State University Ph.D. thesis (2018), https://www.hawc-observatory.org/publications/#thesis.
3. H. Martínez-Huerta and A. Pérez-Lorenzana, Phys. Rev. D **95**, 063001 (2017); H. Martínez-Huerta, these proceedings.
4. Particle Data Group, M. Tanabashi *et al.*, Phys. Rev. D **98**, 030001 (2018), Ch. 39.
5. S. Marinelli, Michigan State University Ph.D. thesis (2019), https://www.hawc-observatory.org/publications/#thesis.
6. V. Vasileiou *et al.*, Phys. Rev. D **87**, 122001 (2013).

Noncommutative Gravity and the Standard-Model Extension

Charles D. Lane

Department of Physics, Berry College, Mount Berry, GA 30149, USA

Indiana University Center for Spacetime Symmetries, Bloomington, IN 47405, USA

Noncommutative geometry has become popular mathematics for describing speculative physics beyond the Standard Model. Noncommutative QED has long been known to fit within the framework of the Standard-Model Extension. We argue in this work that noncommutative gravity also fits within the Standard-Model Extension framework.

The original inspiration for considering noncommutative geometry in physics[1] was the desire to have a Heisenberg-like uncertainty relation for position coordinates: $\Delta x \Delta y > 0$, which corresponds to noncommutativity between position coordinates, $[x, y] \neq 0$. This idea may be made compatible with *observer* Lorentz symmetry by assuming $[x^\mu, x^\nu] = i\theta^{\mu\nu}$, where $\theta^{\mu\nu}$ is real and antisymmetric. (Note that the existence of a nonzero tensor that appears to be a property of spacetime itself violates *particle* Lorentz symmetry.)

A useful tool for constructing noncommutative theories is the Moyal $\star$ product.[2] Consider a commutative field theory with functions/fields $f, g, \ldots$ This may be turned into a noncommutative field theory with noncommutative functions/fields $\hat{f}, \hat{g}, \ldots$ by replacing all ordinary products with $\star$ products:

$$(f \cdot g)(x) \to (\hat{f} \star \hat{g})(x) := \exp\left(\tfrac{i}{2}\theta^{\mu\nu}\tfrac{\partial}{\partial x^\mu}\tfrac{\partial}{\partial y^\nu}\right) \hat{f}(x)\hat{g}(y)\Big|_{x=y}. \qquad (1)$$

Note: (1) This automatically gives $[x^\mu, x^\nu] \to [\hat{x}^\mu, \hat{x}^\nu]_\star = i\theta^{\mu\nu}$ as desired. (2) It has similar form to a multivariable Taylor series, and hence may be related to nonlocality. (3) The Moyal $\star$ product is not the *only* way to define a noncommutative theory; it is simply one convenient approach.

Interpretation of such noncommutative theories is nontrivial as the noncommutative fields $\hat{\psi}, \hat{A}_\mu, \ldots$ do not necessarily correspond to physical particles. A Seiberg–Witten map[3] $\hat{\psi}, \hat{A}_\mu, \ldots \to \psi, A_\mu, \ldots$ is a method of restating noncommutative gauge theories that eases interpretation. This map guarantees that ψ, A_μ are ordinary fields with ordinary gauge transformations whose behavior is physically equivalent to $\widehat{\psi}, \widehat{A}_\mu$.

This strategy has been used to show that noncommutative QED[4] fits within the flat-space SME.[5] In the rest of this work, we relate a model of noncommutative gravity to the gravitational SME.[6]

One way to model gravity is as a spontaneously broken SO(2,3) gauge theory.[7] This provides a good starting place to build a noncommutative model of gravity, as the (broken) gauge symmetry is automatically respected by the Seiberg–Witten map.

The unbroken commutative SO(2,3) action on flat (1+3)-dimensional spacetime may be written $S = c_1 S_1 + c_2 S_2 + c_3 S_3$, where

$$S_1 \sim \mathrm{Tr} \int d^4x \; \varepsilon^{\mu\nu\rho\sigma} F_{\mu\nu} F_{\rho\sigma} \phi, \quad S_2 \sim \mathrm{Tr} \int d^4x \; \varepsilon^{\mu\nu\rho\sigma} F_{\mu\nu} D_\rho \phi D_\sigma \phi \phi,$$

$$\text{and } S_3 \sim \mathrm{Tr} \int d^4x \; \varepsilon^{\mu\nu\rho\sigma} D_\mu \phi D_\nu \phi D_\rho \phi D_\sigma \phi \phi. \tag{2}$$

In this expression, F is the SO(2,3) gauge field, D is the associated covariant derivative, ϕ is a scalar field, and $c_1, \ldots, c_3$ are undetermined weights.

If we then assume that ϕ spontaneously breaks the SO(2,3) symmetry in its ground state, $\langle \phi \rangle = (0,0,0,0,\ell)$, and expand the action around this ground state, then it takes a form that includes conventional gravity: $S \supset -\frac{1}{16\pi G_N} \int d^4x \, e \left(R - \frac{6}{\ell^2}(1 + c_2 + 2c_3) \right)$.

This model may then inspire a noncommutative gravitational theory[8] by following a similar prescription to that followed for noncommutative QED: (1) Start with the unbroken SO(2,3) action. (2) Replace fields F, ϕ with Moyal $\star$ products of noncommutative fields $\widehat{F}, \widehat{\phi}$. (3) Apply a Seiberg–Witten map to replace noncommutative fields with physically equivalent commutative fields. (4) Assume that the SO(2,3)$_\star$ symmetry is spontaneously broken by ϕ having a nonzero vacuum expectation value. The resulting action is left with a noncommutative SO(1,3)$_\star$ symmetry. It may be expanded in powers of $\theta^{\mu\nu}$, taking the form

$$S_{\mathrm{NCR}} = -\int \frac{d^4x \; e}{16\pi G_N} \left\{ \left[R - \frac{6(1+c_2+2c_3)}{\ell^2} \right] + \frac{1}{8\ell^4} \sum_{u=1}^{6} \theta^{\alpha\beta} \theta^{\gamma\delta} C_{(u)} L^{(u)}_{\alpha\beta\gamma\delta} \right\}. \tag{3}$$

The initial bracketed term describes conventional General Relativity. The noncommutative modification is a sum of geometric quantities $L^{(u)}$ and their weights $C_{(u)}$, which are listed in Table 1.

Table 1: Quantities appearing in the noncommutative action.

u	Weight $C_{(u)}$	Geometric Quantity $L^{(u)}_{\alpha\beta\gamma\delta}$
1	$3c_2 + 16c_3$	$R_{\alpha\beta\gamma\delta}$
2	$-6 - 22c_2 - 36c_3$	$g_{\beta\delta}R_{\alpha\gamma}$
3	$\frac{1}{\ell^2}(6 + 28c_2 + 56c_3)$	$g_{\alpha\gamma}g_{\beta\delta}$
4	$-4 - 16c_2 - 32c_3$	$e^{\mu}_a e_{\beta b}(\widetilde{\nabla}_{\gamma}e^a_{\alpha})(\widetilde{\nabla}_{\delta}e^b_{\mu})$
5	$4 + 12c_2 + 32c_3$	$e_{\delta a}e^{\mu}_b(\widetilde{\nabla}_{\gamma}e^a_{\alpha})(\widetilde{\nabla}_{\beta}e^b_{\mu})$
6	$2 + 4c_2 + 8c_3$	$g_{\beta\delta}e^{\mu}_a e^{\nu}_b[(\widetilde{\nabla}_{\alpha}e^a_{\nu})(\widetilde{\nabla}_{\gamma}e^b_{\mu}) - (\widetilde{\nabla}_{\gamma}e^a_{\mu})(\widetilde{\nabla}_{\alpha}e^b_{\nu})]$

The action in Eq. (3) approximately works as a model for noncommutative gravity, though there are some interpretational issues. First, it assumes that $\partial_\alpha\theta^{\mu\nu} = 0$, which is a coordinate-dependent statement. We may try to maintain coordinate independence by requiring that $\nabla_\alpha\theta^{\mu\nu} = 0$. However, such covariant-constant tensors cannot exist in most spacetimes.[9,10]

Second, the derivative $\widetilde{\nabla}$ that appears is covariant with respect to the $SO(1,3)_\star$ connection but not the Christoffel connection: $\widetilde{\nabla}_\gamma e_\alpha{}^a = \partial_\gamma e_\alpha{}^a + \omega_\gamma{}^{ab}e_{ab} = \Gamma^\rho{}_{\gamma\alpha}e_\rho{}^a$. This means that the Christoffel symbols appear explicitly in the action. The troublesome terms where they appear violate observer-diffeomorphism symmetry, though they do respect *local* observer Lorentz transforms. For the rest of this work, we assume that these issues are negligible in experimentally relevant situations. Further, we work at quadratic order in $h_{\mu\nu} = g_{\mu\nu} - \eta_{\mu\nu}$.

To quadratic order in h, the gravitational SME may be written[11] $S_{\text{SME}} \supset \frac{1}{64\pi G_N}\int d^4x\, h_{\mu\nu}\sum_d \widehat{\mathcal{K}}^{(d)\mu\nu\rho\sigma}h_{\rho\sigma}$. The noncommutative action (3) contains many terms of this form,[6] though we only describe a few here:

$$S_{\text{NCR}} \supset \frac{1}{64\pi G_N}\int d^4x\, h_{\mu\nu}\left\{s^{(2,1)\mu\rho\nu\sigma} + s^{(4)\mu\rho\alpha\nu\sigma\beta}\partial_\alpha\partial_\beta + \cdots\right\}h_{\rho\sigma}. \quad (4)$$

First, we may match the $u = 3$ mass-like term in S_{NCR}:

$$S_{\text{NCR,mass}} = \frac{1}{64\pi G_N}\int d^4x\left\{\left[\frac{C_{(3)}}{2\ell^6}\theta^2\right] + \left[\frac{C_{(3)}}{2\ell^6}\left(\tfrac{1}{2}\theta^2\eta^{\rho\sigma} + 2\theta_\alpha{}^\rho\theta^{\alpha\sigma}\right)\right]h_{\rho\sigma}\right.$$
$$\left. + h_{\mu\nu}\left[s^{(2,1)\mu\rho\nu\sigma} + k^{(2,1)\mu\nu\rho\sigma}\right]h_{\rho\sigma}\right\}. \quad (5)$$

The first term is an irrelevant constant, while the 2nd corresponds to a constant stress-energy. The bottom line contains effective values of SME coefficients:

$$s^{(2,1)\mu\rho\nu\sigma} = \frac{C_{(3)}}{12\ell^4}\left[2\eta^{\mu\nu}\theta^{\rho\alpha}\theta^\sigma{}_\alpha + 2\theta^{\rho\nu}\theta^{\sigma\mu} + \cdots\right] \quad \text{and}$$
$$k^{(2,1)\mu\nu\rho\sigma} = \frac{C_{(3)}}{48\ell^4}\left[4\eta^{\mu\nu}\theta^{\rho\alpha}\theta^\sigma{}_\alpha + \cdots\right]. \quad (6)$$

Second, we consider a sample kinetic effect with contributions from the $u =$1, 2, 4, and 5 terms:

$$S_{\text{NCR,kinetic}} \supset \frac{1}{64\pi G_N} \int d^4 x \; h_{\mu\nu} \left\{ s^{(4)\mu\rho\alpha\nu\sigma\beta} \partial_\alpha \partial_\beta + \cdots \right\} h_{\rho\sigma}, \qquad (7)$$

where

$$s^{(4)\mu\rho\alpha\nu\sigma\beta} \sim \frac{2C_{(1)} - 3C_{(2)} + C_{(4)} + C_{(5)}}{\ell^4} \varepsilon^{\mu\rho\alpha\kappa} \varepsilon^{\nu\sigma\beta\lambda} \left[\theta_{\kappa\gamma}\theta_\lambda{}^\gamma - \frac{1}{4}\eta_{\kappa\lambda}\theta^2 \right]. \qquad (8)$$

This coefficient regulates behavior similar to the $\bar{s}^{\mu\nu}$ coefficient that appears in the minimal gravitational SME.[9] We may therefore exploit existing bounds[12] on $\bar{s}^{\mu\nu}$ to extract rough bounds on $\theta^{\mu\nu}$ (albeit bounds that depend on the gauge-breaking scale ℓ):

$$\left| \frac{\theta_{\mu\nu}\theta^{\mu\nu}}{\ell^4} \right| \lesssim 10^{-15}. \qquad (9)$$

Acknowledgments

I would like to thank Berry College and the IU Center for Spacetime Symmetries for support during the creation of this work.

References

1. H.S. Snyder, Phys. Rev. **71**, 38 (1947); A. Connes, *Noncommutative Geometry*, Academic Press, 1994.
2. J.E. Moyal and M.S. Bartlett, Math. Proc. Cam. Phil. Soc. **45**, 99 (1949).
3. N. Seiberg and E. Witten, J. High Energy Phys. **09**, 032 (1999).
4. A.A. Bichl *et al.*, TUW-01-03, UWTHPH-2001-9 (2001); hep-th/0102103.
5. S. Carroll *et al.*, Phys. Rev. Lett. **87**, 141601 (2001).
6. Q.G. Bailey and C.D. Lane, Symmetry **10**, 480 (2018).
7. M.D. Ćirić and V. Radovanović, Phys. Rev. D **89**, 125021 (2014).
8. M.D. Ćirić *et al.*, Phys. Rev. D **96**, 064029 (2017).
9. V.A. Kostelecký, Phys. Rev. D **69**, 105009 (2004).
10. C.D. Lane, Phys. Rev. D **94**, 025016 (2016).
11. V.A. Kostelecký and M. Mewes, Phys. Lett. B **779**, 136 (2018).
12. Q.G. Bailey and V.A. Kostelecký, Phys. Rev. D **74**, 045001 (2006); B.P. Abbott *et al.*, Astrophys. J. **848**, L13 (2017).

Matter-Wave Interferometry for Inertial Sensing and Tests of Fundamental Physics

D. Schlippert, C. Meiners, R.J. Rengelink, C. Schubert, D. Tell, É. Wodey,
K.H. Zipfel, W. Ertmer, and E.M. Rasel

*Institut für Quantenoptik, Leibniz Universität Hannover,
Welfengarten 1, 30167 Hannover, Germany*

Very Long Baseline Atom Interferometry corresponds to ground-based atomic matter-wave interferometry on large scales in space and time, letting the atomic wave functions interfere after free evolution times of several seconds or wave-packet separation at the scale of meters. As inertial sensors, e.g., accelerometers, these devices take advantage of the quadratic scaling of the leading-order phase shift with the free-evolution time to enhance their sensitivity, giving rise to compelling experiments. With shot-noise-limited instabilities better than $10^{-9}\,\mathrm{m/s^2}$ at $1\,\mathrm{s}$ at the horizon, Very Long Baseline Atom Interferometry may compete with state-of-the-art superconducting gravimeters, while providing absolute instead of relative measurements. When operated with several atomic states, isotopes, or species simultaneously, tests of the universality of free fall at a level of parts in 10^{13} and beyond are in reach. Finally, the large spatial extent of the interferometer allows one to probe the limits of coherence at macroscopic scales as well as the interplay of quantum mechanics and gravity. We report on the status of the Very Long Baseline Atom Interferometry facility, its key features, and future prospects in fundamental science.

1. Introduction

Nearly half a century after the seminal observation of gravity-induced phase shifts on matter waves,[1] coherent control in atom interferometers is now a standard technique used around the world to perform inertial measurements,[2] determine fundamental constants,[3,4] and test the laws of fundamental physics.[5] At a given phase noise, the intrinsic sensitivity of an atom accelerometer in the common Mach–Zehnder-type configuration[6] scales with the leading-order phase shift,

$$\Delta\phi = \vec{k}_{\mathrm{eff}} \cdot \vec{a}\, T^2, \tag{1}$$

where $\hbar\vec{k}_{\mathrm{eff}}$ is the momentum transferred during atom–light interaction, $\vec{a}$ is the acting acceleration, and $2T$ is the wave packets' free-evolution time. One can therefore increase the scale factor of the instrument $k_{\mathrm{eff}}T^2$ by exploiting its linear scaling in atomic recoil $\hbar k_{\text{eff}}$ but also the quadratic dependency on the free-evolution time T. To that end, besides initiatives for operation in microgravity,[7-9] increasing the free-fall distance of ground-based devices towards long free-fall tubes is a path pursued by groups around the world.[10,11] Indeed, combined with an exquisite control over external fields and other deteriorating effects, extending the baseline of gravimeters from tens of centimeters to several meters opens the way for competition with state-of-the-art superconducting gravimeters or quantum tests of the universality of free fall at an unprecedented level,[12] competitive with those achieved by the best classical methods.[13-15] In these proceedings, we report on the status and key features of the Very Long Baseline Atom Interferometry (VLBAI) facility implemented at the Hannover Institute of Technology (HITec) of Leibniz Universität Hannover.

2. The VLBAI facility

2.1. *Design*

The VLBAI facility in Hannover is currently in its final stage of construction and consists of three main components:

(1) At the heart of the device is a 10 m long, vertically oriented ultra-high vacuum tube with a 10 cm diameter in clearance for the atom optics light fields. In order to shield the experiment from stray magnetic fields mimicking inertial forces, the tube is enclosed in a high-performance dual-layer magnetic shield reducing magnetic-field gradients to values below 10 nT/m. Additionally, the full baseline is equipped with a network of temperature probes for detecting and subsequently correcting errors due to temperature gradients.[16]

(2) Both ends of the vacuum tube will be equipped with dual-species sources[a] providing the operator with (near-)quantum degenerate ensembles of stable bosonic and fermionic ytterbium isotopes as well as ^{87}Rb.[12] Making use of hybrid magnetic and optical trapping techniques as well as delta-kick collimation,[17] we anticipate an atom flux on the order of 10^6 at/s with temperatures in the picokelvin regime.[18]

(3) For absolute measurements, we will utilize a seismic attenuation system (SAS) to suspend a retro reflection mirror that serves as the atom interferometer's inertial reference. Based on geometric anti-springs,[19]

---

[a]As such, the facility can be operated either in drop mode ($2T = 0.8$ s) or with atoms being launched ($2T = 2.8$ s).

our SAS features a passive resonance frequency of hundreds of milli-hertz and provides means of 6-degrees-of-freedom active stabilization via electromagnetic actuation using the signals of on-board broadband seismometers as well as novel opto-mechanical devices.[20]

The VLBAI facility aims to use rubidium, which is well-established as a standard choice for inertial sensors using readily available source and laser technology, as well as the heavy lanthanide ytterbium, which offers its own benefits. Indeed, as an effective two-electron system it has broad lines to apply strong cooling forces, as well as narrow intercombination transitions with a low Doppler-limit. Furthermore, the bosonic isotopes all share the property of having a vanishing first-order magnetic susceptibility, making systematic effects and environmental decoherence mechanisms easier to control. Finally, the internal composition of our species offers enhanced sensitivity when searching for new physics beyond the Standard Model.[21,22]

2.2. *Performance estimation*

With the SAS and the magnetic shield tackling the dominant external noise sources, we expect an improvement in absolute gravimetry beyond the state of the art. The performance of the device will, however, ultimately be limited by quantum projection, i.e., shot noise. In a simple drop configuration with 2×10^5 at per cycle, 3 s preparation time, and a free-evolution time $2T = 800$ ms, the shot noise-limited sensitivity of the VLBAI facility is 1.7 nm s^{-2} at 1 s. In a more advanced configuration, launching 10^6 at per cycle to reach a free-evolution time of $2T = 2.8$ s and using four-photon atom optics yields a shot-noise-limited acceleration sensitivity of 40 pm s^{-2} at 1 s. For a gradiometric configuration[3,23] with a baseline $L = 5$ m between the two atom interferometers with 10^5 at per cycle, 3 s preparation time, and $2T = 800$ ms, the shot-noise-limited sensitivity is 5×10^{-10} s^{-1} at 1 s.

A simultaneous-comparison measurement with ^{87}Rb and ^{170}Yb has the perspective for testing the universality of free fall[12] by determining the Eötvös ratio between the two species to better than one part in 10^{13}.

2.3. *Fundamental physics*

Beyond these metrological goals, the device will also serve as a test bed for interferometry with very large scale factors as necessary for gravitational-wave detection[23] and is intrinsically sensitive to fundamental decoherence mechanisms[24] and limitations of the quantum superposition principle.[25]

3. Conclusion and outlook

The VLBAI facility will enable highly sensitive absolute gravimetry and tests of the universality of free fall. It therefore offers a world-wide unique environment, both for cutting-edge inertial sensing and to test our understanding of General Relativity, quantum mechanics, and their interplay, possibly leading to a future reconciliation of the two.[12]

Acknowledgments

This project is supported through the CRCs 1128 "geo-Q" and 1227 "DQ-mat," "Niedersächsisches Vorab" in the "Quantum- and Nano-Metrology (QUANOMET)" initiative. DS gratefully acknowledges funding by the Federal Ministry of Education and Research (BMBF) through the funding program Photonics Research Germany under contract No. 13N14875.

References

1. R. Colella *et al.*, Phys. Rev. Lett. **34**, 1472 (1975).
2. A. Peters, K.Y. Chung, and S. Chu, Nature **400**, 849 (1999).
3. G. Rosi *et al.*, Nature **510**, 518 (2014).
4. R.H. Parker *et al.*, Science **360**, 191 (2018).
5. M. Jaffe *et al.*, Nat. Phys. **13**, 938 (2017); D.O. Sabulsky *et al.*, Phys. Rev. Lett. **123**, 061102 (2019).
6. M. Kasevich and S. Chu, Phys. Rev. Lett. **67**, 181 (1991).
7. D.N. Aguilera *et al.*, Class. Quant. Grav. **31**, 115010 (2014).
8. D. Becker *et al.*, Nature **562**, 391 (2018).
9. L. Wörner *et al.*, in *Proceedings of the 68th International Astronautic Congress 2018*, IAC-18,A2,1,13,x45004 (2018).
10. C. Overstreet *et al.*, Phys. Rev. Lett. **120**, 183604 (2018).
11. L. Zhou *et al.*, Gen. Rel. Grav. **43**, 1931 (2011).
12. J. Hartwig *et al.*, New J. Phys. **17**, 035011 (2015).
13. P. Touboul *et al.*, Phys. Rev. Lett. **119**, 231101 (2017).
14. F. Hofmann and J. Müller, Class. Quant. Grav. **35**, 035015 (2018).
15. S. Schlamminger *et al.*, Phys. Rev. Lett. **100**, 041101 (2008).
16. P. Haslinger *et al.*, Nat. Phys. **14**, 1745 (2017).
17. H. Müntinga *et al.*, Phys. Rev. Lett. **110**, 093602 (2013).
18. S. Loriani *et al.*, New J. Phys. **21**, 063030 (2019).
19. A. Bertolini *et al.*, Nucl. Instrum. Meth. A **435**, 475 (1999).
20. L.L. Richardson *et al.*, arXiv:1902.02867.
21. M.A. Hohensee *et al.*, Phys. Rev. Lett. **111**, 151102 (2013).
22. V.A. Kostelecký and J.D. Tasson, Phys. Rev. D **83**, 016013 (2011).
23. S. Dimopoulos *et al.*, Phys. Rev. D **78**, 122002 (2008).
24. A. Bassi *et al.*, Class. Quant. Grav. **34**, 193002 (2017).
25. T. Kovachy *et al.*, Nature **528**, 530 (2015).

Developments in Lorentz and CPT Violation

V. Alan Kostelecký

Physics Department, Indiana University, Bloomington, IN 47405, USA

This talk at the CPT'19 meeting outlines a few recent developments in Lorentz and CPT violation, with particular attention to results obtained by researchers at the Indiana University Center for Spacetime Symmetries.

1. Introduction

Motivated by the prospect of minuscule observable effects arising from Planck-scale physics, searches for Lorentz and CPT violation have made impressive advances in recent years.[1] The scope of ongoing efforts presented at the CPT'19 meeting indicates that this rapid pace of development will continue unabated, with experiments achieving sensitivities to Lorentz violation that are orders of magnitude beyond present capabilities and providing unprecedented probes of the CPT theorem. Substantial theoretical advances in the subject are also being made, and the prospects are excellent for completing a comprehensive description of possible effects on all forces and particles and for achieving a broad understanding of the underlying mathematical structure in the near future. In this talk, I summarize some basics of the subject and outline a few recent results obtained at the Indiana University Center for Spacetime Symmetries (IUCSS).

2. Basics

No compelling experimental evidence for Lorentz or CPT violation has been reported to date, so any effects are expected to involve only tiny deviations from the physics of General Relativity (GR) and the Standard Model (SM). In studying the subject, it is therefore desirable to work within a theoretical description of Lorentz and CPT violation that is both model independent and includes all possibilities consistent with the structure of GR and the SM. The natural context for a description of this type is effective field theory.[2] The realistic and coordinate-independent effective field theory for Lorentz and CPT violation is known as the Standard-Model Extension

(SME).[3,4] It can be obtained by incorporating all coordinate-independent and Lorentz-violating terms in the action for GR coupled to the SM. These terms also describe general realistic CPT violation,[3,5] and they are compatible with either spontaneous or explicit Lorentz violation in an underlying unified theory such as strings.[6]

The SME action incorporates Lorentz-violating operators of any mass dimension d, with the minimal SME defined to include the subset of operators of renormalizable dimension $d \leq 4$. A given SME term is formed as the observer-scalar contraction of a Lorentz-violating operator with a coefficient for Lorentz violation that acts as a background coupling to control observable effects. The propagation and interactions of each species are modified and can vary with momentum, spin, and flavor. All minimal-SME terms[3,4] and many nonminimal terms[7–9] have been explicitly constructed. The resulting experimental signals are expected to be suppressed either directly or through a mechanism such as countershading via naturally small couplings.[10] Impressive constraints on SME coefficients from many experiments have been obtained.[1] The generality of the SME framework insures these constraints apply to any specific Lorentz-violating model that is consistent with realistic effective field theory.

The geometry of Lorentz violation is an interesting issue for exploration. If the Lorentz violation is spontaneous, then the geometry can remain Riemann or Riemann–Cartan[4] and the phenomenology incorporates Nambu–Goldstone modes.[11] However, if the Lorentz violation is explicit, then the geometry cannot typically be Riemann and is conjectured to be Finsler instead.[4] Support for this idea has grown in recent years,[12,13] but a complete demonstration is lacking at present.

3. Developments from the IUCSS

In the past three years, developments from the IUCSS have primarily involved the quark, gauge, and gravity sectors. In the quark sector, direct constraints on minimal-SME coefficients can be extracted using neutral-meson oscillations,[14] and numerous experiments on K, D, B_d, and B_s mixing have achieved high sensitivities to CPT-odd effects on the u, d, s, c, and b quarks.[15] Recent work reveals that nonminimal quark coefficients at $d = 5$ provide numerous independent measures of CPT violation,[16] many of which are experimentally unconstrained to date. The t quark decays too rapidly to hadronize, but t–$\bar{t}$ pair production and single-t production are sensitive to t-sector coefficients and are the subject of ongoing experimental analyses.[17] High-energy studies of deep inelastic scattering and Drell-Yan processes also offer access to quark-sector coefficients,[18] and corresponding data analyses are being pursued. Another active line of reasoning adapts chiral perturbation theory to relate quark coefficients to hadron coefficients,[19] yielding further tests of Lorentz and CPT symmetry.

In the gauge sector, the long-standing challenge of constructing all nonabelian Lorentz-violating operators at arbitrary d has recently been solved.[8] The methodology yields all matter–gauge couplings, so the full Lorentz- and CPT-violating actions for quantum electrodynamics, quantum chromodynamics, and related theories are now available for exploration. Constraints on photon-sector coefficients continue to improve.[20] Signals of Lorentz violation arising in clock-comparison experiments at arbitrary d have recently been studied,[21] revealing complementary sensitivities from fountain clocks, comagnetometers, ion traps, lattice clocks, entangled states, and antimatter. These various advances suggest excellent prospects for future searches for Lorentz and CPT violation in the gauge and matter sectors.

In the gravity sector, all operators modifying the propagation of the metric perturbation $h_{\mu\nu}$, including ones preserving or violating Lorentz and gauge invariance, have been classified and constructed.[9] Many of the corresponding coefficients are unexplored but could be measured via gravitation-wave and astrophysical observations. A general methodology exists for analyzing Lorentz-violation searches in experiments on short-range gravity,[9] and constraints on certain coefficients with d up to eight have now been obtained.[22] Work in progress further extends gravity-sector tests to matter–gravity couplings at arbitrary d.[23] Results from the SME can also be applied to constrain hypothesized Lorentz-invariant effects whenever these lead to nonzero background values for vector or tensor objects. This idea recently yielded the first experimental constraints on all components of nonmetricity.[24] At the foundational level, further confirmation of the correspondence between the SME and Finsler geometry has been established via the construction of all Finsler geometries for spin-independent Lorentz-violating effects.[13] The scope and breadth of all these results augurs well for future advances in the gravity sector on both theoretical and experimental fronts.

Acknowledgments

This work was supported in part by US DOE grant DE-SC0010120 and by the Indiana University Center for Spacetime Symmetries.

44

References

1. V.A. Kostelecký and N. Russell, arXiv:0801.0287v12.
2. See, e.g., S. Weinberg, Proc. Sci. CD **09**, 001 (2009).
3. D. Colladay and V.A. Kostelecký, Phys. Rev. D **55**, 6760 (1997); Phys. Rev. D **58**, 116002 (1998).
4. V.A. Kostelecký, Phys. Rev. D **69**, 105009 (2004).
5. O.W. Greenberg, Phys. Rev. Lett. **89**, 231602 (2002).
6. V.A. Kostelecký and S. Samuel, Phys. Rev. D **39**, 683 (1989); V.A. Kostelecký and R. Potting, Nucl. Phys. B **359**, 545 (1991); Phys. Rev. D **51**, 3923 (1995); V.A. Kostelecký and R. Lehnert, Phys. Rev. D **63**, 065008 (2001).
7. V.A. Kostelecký and M. Mewes, Phys. Rev. D **80**, 015020 (2009); Phys. Rev. D **85**, 096005 (2012); Phys. Rev. D **88**, 096006 (2013); Y. Ding and V.A. Kostelecký, Phys. Rev. D **94**, 056008 (2016).
8. V.A. Kostelecký and Z. Li, Phys. Rev. D **99**, 056016 (2019).
9. Q.G. Bailey *et al.*, Phys. Rev. D **91**, 022006 (2015); V.A. Kostelecký and M. Mewes, Phys. Lett. B **757**, 510 (2016); Phys. Lett. B **766**, 137 (2017); Phys. Lett. B **779**, 136 (2018).
10. V.A. Kostelecký and J.D. Tasson, Phys. Rev. Lett. **102**, 010402 (2009); Phys. Rev. D **83**, 016013 (2011).
11. R. Bluhm and V.A. Kostelecký, Phys. Rev. D **71**, 065008 (2005); V.A. Kostelecký and R. Potting, Gen. Rel. Grav. **37**, 1675 (2005); Phys. Rev. D **79**, 065018 (2009); B. Altschul *et al.*, Phys. Rev. D **81**, 065028 (2010).
12. M. Schreck, Phys. Lett. B **793**, 70 (2019); D. Colladay and P. McDonald, Phys. Rev. D **85**, 044042 (2012); V.A. Kostelecký, Phys. Lett. B **701**, 137 (2011); V.A. Kostelecký and N. Russell, Phys. Lett. B **693**, 2010 (2010).
13. B.R. Edwards and V.A. Kostelecký, Phys. Lett. B **786**, 319 (2018).
14. V.A. Kostelecký, Phys. Rev. Lett. **80**, 1818 (1998).
15. K.R. Schubert, arXiv:1607.05882; R. Aaij *et al.*, Phys. Rev. Lett. **116**, 241601 (2016); V.M. Abazov *et al.*, Phys. Rev. Lett. **115** 161601 (2015); D. Babusci *et al.*, Phys. Lett. B **730**, 89 (2014).
16. B.R. Edwards and V.A. Kostelecký, Phys. Lett. B **795**, 620 (2019).
17. V.M. Abazov *et al.*, Phys. Rev. Lett. **108**, 261603 (2012); M.S. Berger *et al.*, Phys. Rev. D **93**, 036005 (2016); A. Carle *et al.*, arXiv:1908.11256.
18. V.A. Kostelecký *et al.*, Phys. Lett. B **769**, 272 (2017); in preparation.
19. R. Kamand *et al.*, Phys. Rev. D **95**, 0556005 (2017); Phys. Rev. D **97**, 095027 (2018); B. Altschul and M.R. Schindler, arXiv:1907.02490; J.P. Noordmans *et al.*, Phys. Rev. C **94**, 025502 (2016).
20. L. Pogosian *et al.*, Phys. Rev. D **100**, 023407 (2019); F. Kislat, Symmetry **10**, 596 (2018); F. Kislat and H. Krawczynski, Phys. Rev. D **95**, 083013 (2017); J.J. Wei *et al.*, Ap. J. **842**, 115 (2017).
21. V.A. Kostelecký and A.J. Vargas, Phys. Rev. D **98**, 036003 (2018).
22. C.G. Shao *et al.*, Phys. Rev. Lett. **122**, 011102 (2019); Phys. Rev. Lett. **117**, 071102 (2016).
23. V.A. Kostelecký and Z. Li, in preparation.
24. J. Foster *et al.*, Phys. Rev. D **95**, 084033 (2017).

Prospects for Lorentz-Violation Searches at the LHC and Future Colliders

N. Chanon, A. Carle, and S. Perriès

Université de Lyon, Université Claude Bernard Lyon 1,
CNRS-IN2P3, Institut de Physique Nucléaire de Lyon,
Villeurbanne 69622, France

Hadron colliders are providing a unique opportunity for testing Lorentz invariance and CPT symmetry at high energy and in a laboratory. A first measurement in the top-quark sector was performed at the Tevatron. We present here prospective studies for testing Lorentz invariance in top-quark pair production at the LHC and future colliders. The b-quark sector was investigated recently at LHCb. Eventually, new bounds on photon parameters can be extracted from the observation of TeV photons at the LHC. We will conclude by highlighting other opportunities provided by hadron colliders.

1. Testing Lorentz invariance at hadron colliders

Lorentz invariance is a fundamental symmetry of the Standard Model, not necessarily expected to be conserved in theories of quantum gravity, such as string theories.[1] Hadron colliders offer the opportunity to probe Lorentz invariance and CPT symmetry (CPT breaking implies violation of Lorentz invariance under mild assumptions) with elementary particles at high energy. The LHC provides proton–proton collisions at a center-of-mass energy of 13 TeV (Run 2). Although the scale of Lorentz- or CPT-symmetry breaking is sometimes expected to lie at the Planck mass, the extra-dimensions paradigm could lower the scale of quantum gravity. Possible remnants of Lorentz-symmetry breaking can be analyzed using signatures proposed within the framework of an effective field theory, the Standard-Model Extension (SME).[2] Among the processes occurring with the highest cross sections in p–p collisions, and where such signatures could be searched for, one can list jet production, photon production, W or Z production, and $t\bar{t}$ production. We will focus in these proceedings on photon and top-quark production as a probes for Lorentz-violation (LV) searches.

2. Probing LV in the top-quark sector at hadron colliders

2.1. *Searches for LV with $t\bar{t}$ production*

The top-quark sector in the SME is weakly constrained with only one direct measurement, performed using the D0 detector at the Tevatron[3] via observation of $t\bar{t}$ events. The obtained results are compatible with no LV with an absolute uncertainty of 10%.

The top-quark lagrangian in the SME contains both CPT-even and CPT-odd LV contributions. Since top and antitop CPT-violating corrections cancel in $t\bar{t}$ production, we will focus on the CPT-even LV contribution. The SME introduces the LV $c_{\mu\nu}$ coefficients, modifying the top-quark propagator, the Wtb vertex and the top–gluon coupling. The squared matrix elements for top-quark production and decay including these corrections are known[4] at leading order in perturbative QCD.

The LHC is a top-quark factory. From the Tevatron to the LHC, the center-of-mass energy has increased from 1.96 TeV to 13 TeV. The cross section for $t\bar{t}$ production has increased by a factor $\approx$115 owing to the gluon luminosity in the proton. Furthermore, the recorded luminosity at LHC Run 2 is about 150 fb^{-1}, while D0 results[3] were obtained analyzing 5.3 fb^{-1}. An even higher number of $t\bar{t}$ events will be produced at future hadron colliders, such as the HL-LHC (3 ab^{-1} at 14 TeV) and the HE-LHC or FCC-hh options (15 ab^{-1} at 27 TeV and 100 TeV, respectively).

We compute the change in $t\bar{t}$ cross section in the SME framework. The SME coefficients are constant in a given inertial reference frame taken by convention to be the Sun-centered frame. The rotation of the Earth around its axis induces a sinusoidal modulation of the $t\bar{t}$ cross section as a function of time. We use the location of ATLAS or CMS experiments (both detectors are located at opposite azimuth in the LHC ring and "see" the same cross section). The experiments are sensitive to the same $c_{\mu\nu}$ coefficients as those measured at D0. It is found that the amplitude of the induced modulation is growing with the center-of-mass energy.

We evaluate the expected precision on $c_{\mu\nu}$ for several benchmarks: D0, LHC Run 2, HL-LHC, HE-LHC, FCC.[5] From an absolute precision of 0.1 on the coefficients measured at D0, we find that the expected sensitivity would improve by a factor 10^2–10^3 at LHC Run 2, the exact value depending on the $c_{\mu\nu}$ benchmark. An additional improvement by another factor 10^2 is expected at future colliders like the FCC. As a summary, there is great potential for precision measurement of top-quark $c_{\mu\nu}$ coefficients at present and future hadron colliders.

2.2. *Searches for CPT violation with single-top production*

The single-top-quark production is sensitive to CPT violation. By comparing the rates for single-top and single-antitop production as a function of time, the b_μ coefficient in the SME could be measured.[4] However, the measurement is challenging owing to a huge $t\bar{t}$ background. While the s-channel single-top production remains to be observed at the LHC, the t-channel or tW-channel could be investigated.

3. A test of CPT violation through B_s oscillations at LHCb

The first test of CPT violation at the LHC in the context of the SME was recently performed at LHCb using B^0 and B_s neutral-meson oscillations.[6] The tiny mass difference between the B_s and its antiparticle is used to achieve excellent precision on the Δa_μ SME coefficients for b quarks. The analysis, performed as a function of sidereal time, achieved a precision of 10^{-14} GeV, a factor 10^2 improvement relative to previous measurements.

While this measurement was performed with the LHC data collected at 7 and 8 TeV, it could already be improved by using the 13 TeV data of LHC Run 2. There is also potential for improvement by a factor of three of the limits on c-quark SME coefficients by analyzing the $D^0 \to K^- \pi^+$ decay.[7]

4. Direct photons and constraints on the SME

LV in the photon sector can, for example, be searched for with vacuum birefringence, polarization-independent anisotropies of the phase-speed of light, or an isotropic shift in the phase-speed of light.[8] The latter is controlled by the $\tilde{\kappa}_{tr}$ coefficient in the SME. A positive $\tilde{\kappa}_{tr}$ leads to Cherenkov radiation of charged particles in vacuum, and a negative $\tilde{\kappa}_{tr}$ leads to photon decay into a fermion–antifermion pair in vacuum; both of these processes are forbidden in the Standard Model. The smaller $-\tilde{\kappa}_{tr}$, the higher the threshold energy at which photon decay occurs, and the longer photons travel before decaying. The interplay between $\tilde{\kappa}_{tr}$ and the fermion $c_{\mu\nu}$ coefficients implies that only the quantity $\tilde{\kappa}_{tr} - (4/3)c_{TT}$ can actually be measured.

Today's best limits on this coefficient are obtained using photons of astrophysical origin. Measurements of high-energy photon showers on Earth imply that the photon traveled a large distance without decaying, hence the threshold for photon decay was not reached, and the photon must have been below the LV pair-production threshold. Bounds on $\tilde{\kappa}_{tr} - (4/3)c_{TT}$ can then be extracted (assuming, for instance, electrons as decay products).

The D0 experiment at the Tevatron observed photons up to $340.5\,\text{GeV}$. By assuming conservatively that the threshold for photon decay is $300\,\text{GeV}$, and assuming the process $\gamma \to e^+e^-$, a bound from direct photon production at hadron colliders was set:[8] $\tilde{\kappa}_{tr} - (4/3)c_{TT} > -5.8 \times 10^{-12}$.

At the LHC, the ATLAS experiment recorded photons with a transverse energy above $1.1\,\text{TeV}$ with a minimum pseudorapidity $|\eta| > 0.6$,[9] such that we can derive a threshold energy of $1.3\,\text{TeV}$. A new bound can then be extracted: $\tilde{\kappa}_{tr} - (4/3)c_{TT} > -3.1 \times 10^{-13}$, an improvement by almost a factor 20 relative to the previous collider bound. However, since the energy threshold is higher, the photon would travel on average a longer distance than at D0 before decaying, and in some cases the decay could still be reconstructed as a converted photon. This idea would deserve a more detailed analysis.

5. Conclusions

In these proceedings, we presented an overview of possible signatures for violations of Lorentz invariance and CPT symmetry at hadron colliders involving top quarks, neutral-meson oscillation, and direct photon production. Hadron colliders present also other opportunities: with the high cross section for QCD jet production, or W and Z production, interesting new searches could be performed.

Acknowledgments

The author is thankful to the organizers for the invitation and for the fruitful discussions at the meeting.

References

1. V.A. Kostelecký and S. Samuel, Phys. Rev. D **39**, 683 (1989).
2. D. Colladay and V.A. Kostelecký, Phys. Rev. D **55**, 6760 (1997).
3. D0 Collaboration, V.M. Abazov *et al.*, Phys. Rev. Lett. **108**, 261603 (2012).
4. M.S. Berger, V.A. Kostelecký, and Z. Liu, Phys. Rev. D **93**, 036005 (2016).
5. A. Carle, N. Chanon, and S. Perriès, to appear.
6. LHCb Collaboration, R. Aaij *et al.*, Phys. Rev. Lett. **116**, 241601 (2016).
7. J. van Tilburg and M. van Veghel, Phys. Lett. B **742**, 236 (2015).
8. M.A. Hohensee, R. Lehnert, D.F. Phillips, and R.L. Walsworth, Phys. Rev. D **80**, 036010 (2009).
9. ATLAS Collaboration, M. Aaboud *et al.*, Phys. Lett. B **770**, 473 (2017).

Recent Progress Probing Lorentz Violation at HUST

Ya-Fen Chen, Yu-Jie Tan, and Cheng-Gang Shao

MOE Key Laboratory of Fundamental Physical Quantities Measurements,
Hubei Key Laboratory of Gravitation and Quantum Physics,
PGMF and School of Physics, Huazhong University of Science and Technology,
Wuhan 430074, China

This work mainly discusses recent experimental progress probing the effects of Lorentz violation at $d = 6$ with a special striped-structure experiment to increase the signal of Lorentz breaking. We also propose a new experimental design using the striped geometry with triplex modulation to constrain independently 14 Lorentz-violating coefficients with a higher sensitivity.

1. Introduction

Gravitational phenomena are well described by General Relativity (GR), in which the Einstein Equivalence Principle represents an important foundation. Since the breakdown of local Lorentz invariance may be incompatible with the Equivalence Principle, the study of Lorentz-violation (LV) effects in the spacetime theory of gravity is a new way to explore GR. In the past, limits on LV coefficients in pure gravity have often been obtained by extracting the violation signal from some short-range gravitational experimental data, such as testing the gravitational inverse-square law using a torsion pendulum.[1-3] Recently, we have proposed a torsion-scale experiment with stripe structure to test LV effects,[4,5] and this experiment is ongoing. In this work, we mainly present our progress in the context of this stripe-structure experiment.

2. Pure-gravity LV involving terms quadratic in curvature

As an effective field theory, the Standard-Model Extension (SME) provides a versatile tool to study the effects of LV in low-energy experiments. Its pure-gravity sector can be formulated as a Lagrange density composed of the usual Einstein–Hilbert term R and a cosmological constant; LV terms are expressed as an infinite series of operators with increasing mass dimension $d,$[6]

$$L_d = \frac{\sqrt{-g}}{16\pi G}(R + \Lambda + L_M + L_{LV}^{(4)} + L_{LV}^{(5)} + L_{LV}^{(6)} + \cdots). \quad (1)$$

Here, the last three terms are the LV terms, which are copmposed of a LV coefficient field $k_{\alpha\beta\ldots}$ contracted with curvature quantities. We only focus on the term $L_{LV}^{(6)}$ to carry out our research. In a spherical-coordinate representation, the LV correction to the potential between two masses can be expressed as:

$$V_{LV}(\vec{r}) = -G \sum_{jm} \frac{m_1 m_2}{r^3} Y_{jm}(\theta, \phi) k_{jm}^{lab}, \quad (2)$$

where k_{jm}^{lab} are the LV coefficients expressed in a spherical basis in the lab frame. The goal of the present experiment is to determine limits on these LV coefficients.

3. Studying LV in short-range gravitational experiments

As the SME coefficients k_{jm}^{lab} do not contain scalar pieces, they are frame dependent. The standard convention is to express them in the Sun-centered frame as $k_{jm}^{N(6)}$, where they are taken as constant. As the transformation between the laboratory and Sun frames is time dependent due to the rotation of the Earth, many components of k_{jm}^{lab}, and thus measurable quantities, will vary with sidereal time. It is then convenient to decompose such quantities into a Fourier series with Fourier coefficients determined by $k_{jm}^{N(6)}$. Our observable is torque, and its LV contribution can be expressed as

$$\tau_{LV} = C_0 + \sum_{m=1}^{4} [C_m \cos(m\omega_\oplus T) + S_m \sin(m\omega_\oplus T)]. \quad (3)$$

The detailed relationship between the Fourier amplitudes and the LV coefficients in Eq. (3) has been described in Ref. 5.

Since the LV signal is primarily determined by edge effects, we designed two stripe-type experiments (horizontal and vertical stripe-type geometries) to increase this signal. Through a more in-depth analysis, we also found that placing the experimental setup at a certain azimuth can further enlarge the LV signal. Therefore, we can use angle modulation to jointly analyze the size of the LV effects. Our most direct idea is to put the experimental device on a turntable and rotate the turntable frequency, so that the limit on the LV coefficients can be extracted from the triple modulation in the experimental data. The simplified design for this new experimental setup is shown in Fig. 1.

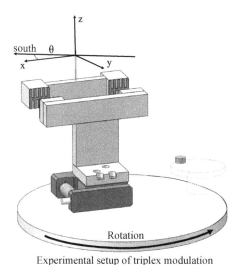

Experimental setup of triplex modulation

Fig. 1: Schematic drawing of the experimental setup for striped-geometry experiments with triplex modulation.

Thus far, we have successfully finished the machining, including the stripe-shape tungsten sheet and a batch of high-precision glass blocks. Figure 2 shows tungsten in panel a) and the glass parts in panel b). All of these items have been measured, and we have verified that their various geometric indices satisfy our targeted accuracy requirements. As the next step, we will be conducting a high-precision bonding assembly.

4. Conclusion

In this work, we mainly reviewed our previous efforts searching for LV effects in the SME's pure-gravity sector in our laboratory. Based on new considerations, we propose a novel experimental design aimed at searching for LV signals that upgrades dual to triplex modulation. In this design, a rotation stage is used to rotate the whole experimental setup to modulate the LV signal. This improvement permits access to all 14 anisotropic SME coefficients. With the processing of parts completed and component assembly and measurement imminent, this new effort is progressing well.

Acknowledgments

This work was supported by the National Natural Science Foundation of China under grant Nos. 91636221 and 11805074.

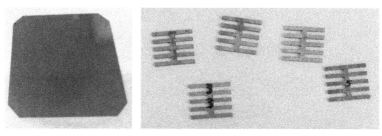

a) Tungsten shape

b) Glass parts

Fig. 2: Fabricated components of the experimental apparatus. Panel a) shows the tungsten sheet and the fringe structure for the experiment, and panel b) displays samples of the glass blocks required for the measurement.

References

1. S.Q. Yang, B.F. Zhan, Q.L. Wang, C.G. Shao, L.C. Tu, W.H. Tan, and J. Luo, Phys. Rev. Lett. **108**, 081101 (2012).
2. J.C. Long and V.A. Kostelecký, Phys. Rev. D **91**, 092003 (2015).
3. W.H. Tan, S.Q. Yang, C.G. Shao, J. Li, A.B. Du, B.F. Zhan, Q.L. Wang, P.S. Luo, L.C. Tu, and J. Luo, Phys. Rev. Lett. **116**, 131101 (2016).
4. C.G. Shao, Y.F. Chen, Y.J. Tan, J. Luo, S.Q. Yang, and M.E. Tobar, Phys. Rev. D **94**, 104061 (2016).
5. Y.F. Chen, Y.J. Tan, and C.G. Shao, Symmetry **9**, 217 (2017).
6. V.A. Kostelecký, Phys. Rev. D **69**, 105009 (2009).

On CPT Tests with Entangled Neutral Kaons

Antonio Di Domenico

Dipartimento di Fisica, Sapienza Università di Roma and INFN Sezione di Roma,
P.le Aldo Moro 2, I-00185, Rome, Italy

Neutral K mesons produced in entangled pairs at a ϕ factory constitute a unique system for the study of CPT symmetry at the utmost sensitivity. In this paper, some aspects of the subtle role of entanglement in the definition of some CPT observables are discussed.

1. The ω effect and semileptonic decays of K_S

At a ϕ factory, neutral-kaon pairs are produced in a pure antisymmetric entangled state:

$$|i\rangle = \frac{1}{\sqrt{2}}[|K^0\rangle|\bar{K}^0\rangle - |\bar{K}^0\rangle|K^0\rangle], \tag{1}$$

making possible neutral-kaon interferometry and offering several peculiar possibilities to study discrete symmetries and the basic principles of quantum mechanics.[1]

In a quantum-gravity framework inducing decoherence, the CPT operator is *ill-defined*. The resulting loss of particle–antiparticle identity might induce a breakdown of the correlation imposed by Bose statistics resulting in the addition of a symmetric component to the initial state (1):

$$|i\rangle \propto |K^0\rangle|\bar{K}^0\rangle - |\bar{K}^0\rangle|K^0\rangle + \omega \left(|K^0\rangle|\bar{K}^0\rangle + |\bar{K}^0\rangle|K^0\rangle\right),$$
$$\propto |K_S\rangle|K_L\rangle - |K_L\rangle|K_S\rangle + \omega \left(|K_S\rangle|K_S\rangle - |K_L\rangle|K_L\rangle\right), \tag{2}$$

where ω is a complex parameter describing this novel CPT-violation phenomenon[2] (termed from now on ω *effect*) that in a quantum-gravity scenario is presumed to be at most of order $|\omega| \sim \left[(m_K^2/M_{\text{Planck}})/\Delta\Gamma\right]^{1/2} \sim 10^{-3}$, with $\Delta\Gamma = \Gamma_S - \Gamma_L$. The first measurement of the ω parameter has been performed by the KLOE collaboration[3,4] analyzing the time difference of the two kaon decays in the CP-violating channel $\phi \to K^0\bar{K}^0 \to$

$\pi^+\pi^-, \pi^+\pi^-$:

$$\mathrm{Re}(\omega) = \left(-1.6^{+3.0}_{-2.1}\mathrm{stat} \pm 0.4\mathrm{syst}\right) \times 10^{-4},$$
$$\mathrm{Im}(\omega) = \left(-1.7^{+3.3}_{-3.0}\mathrm{stat} \pm 1.2\mathrm{syst}\right) \times 10^{-4}, \qquad (3)$$

with a remarkable level of precision and already in the interesting Planck's region.

The KLOE-2 collaboration more recently measured the semileptonic charge asymmetry A_S of the K_S by tagging a pure K_S beam with the detection of a kaon at large times $(t \gg \tau_S)$:[5]

$$A_S = \frac{\Gamma(K_S \to \pi^- e^+ \nu) - \Gamma(K_S \to \pi^+ e^- \bar{\nu})}{\Gamma(K_S \to \pi^- e^+ \nu) + \Gamma(K_S \to \pi^+ e^- \bar{\nu})}$$
$$= (-3.8 \pm 5.0_{\mathrm{stat}} \pm 2.6_{\mathrm{syst}}) \times 10^{-3}. \qquad (4)$$

In the presence of a possible ω effect, the tagged beam is no longer a pure K_S beam due—at first order—to the interference of the $K_L K_L$ term with the leading one, and consequently the observable asymmetry is modified as follows:

$$A_S(\omega) \simeq A_S - \frac{2\Gamma_S}{\sqrt{\Delta m^2 + \frac{(\Gamma_S + \Gamma_L)^2}{4}}} \left[\mathrm{Re}(\omega) \cos \tilde{\phi}_{\mathrm{SW}} + \mathrm{Im}(\omega) \sin \tilde{\phi}_{\mathrm{SW}}\right], \qquad (5)$$

with $\tan \tilde{\phi}_{\mathrm{SW}} = 2\Delta m/(\Gamma_S + \Gamma_L)$. The quantity $\Delta A = A_S(\omega) - A_L$ remains a pure CPT-violation observable,[6] and its measurement implies a linear constraint on the ω effect in the $(\mathrm{Re}\,\omega, \mathrm{Im}\,\omega)$ plane, which, however, results in a rather loose bound due to the limited precision of A_S in (4) and is perfectly compatible with the direct measurement (3).

2. CPT violation and Lorentz-symmetry breaking

In the Standard-Model Extension (SME),[7] CPT violation in the neutral-kaon system manifests itself to lowest order only in the mixing parameter δ and exhibits a dependence on the four-momentum of the kaon:[8]

$$\delta \approx i \sin \phi_{\mathrm{SW}} e^{i\phi_{\mathrm{SW}}} \gamma_K (\Delta a_0 - \vec{\beta}_K \cdot \Delta\vec{a})/\Delta m, \qquad (6)$$

where γ_K and $\vec{\beta}_K$ are the kaon boost factor and velocity in the observer frame, ϕ_{SW} is the so called *superweak* phase, and Δa_μ are four CPT- and Lorentz-violating coefficients for the two valence quarks in the kaon.

As discussed in Refs. 1,9, at a ϕ factory the four Δa_μ parameters can be evaluated using two main alternative methods: (i) by measuring the difference of semileptonic charge asymmetries $\Delta A = A_S - A_L$, or (ii) by studying the interference pattern of the $\phi \to K^0 \bar{K}^0 \to \pi^+\pi^-, \pi^+\pi^-$ final state as a function of sidereal time and particle direction in celestial coordinates.[10]

In this context, as the two entangled kaons in state (1) fly apart in different directions, they may experience different CPT violation, and therefore—as suggested in Refs. 11,12—they evolve in time according to different physical $K_{S,L}$ states, which are no longer identical for the two sides. Introducing the index $i = 1, 2$ to distinguish the two sides, the initial entangled state (1) can be rewritten (at first order in small parameters) as:

$$|i\rangle \propto |K_S^{(1)}\rangle|K_L^{(2)}\rangle - |K_L^{(1)}\rangle|K_S^{(2)}\rangle$$
$$+ \Delta\delta \left(|K_S^{(1)}\rangle|K_S^{(2)}\rangle + |K_L^{(1)}\rangle|K_L^{(2)}\rangle \right), \qquad (7)$$

with $\Delta\delta = \delta^{(2)} - \delta^{(1)}$. It is worth noting that the same relative sign of the $K_S K_S$ and $K_L K_L$ terms in (7) reflects the perfect antisymmetry of the state in the $K^0 \bar{K}^0$ basis, contrary to the case of the ω effect in (2).

As a consequence of the state (7), CPT violation in SME can mimic the ω effect. In case (i), the effect can be easily quantified by Eq. (5) with the substitution $\omega = -\Delta\delta$, while in case (ii) the effect results at first order only from the $K_S K_S$ term (the substitution $\omega = \Delta\delta$ has to be used in this case) and appears negligible.

3. Conclusions

The KLOE experiment at the DAΦNE ϕ factory performed some of the most stringent tests on the validity of CPT symmetry. Some aspects of the subtle role played by entanglement in some of these tests have been briefly discussed in this paper. These considerations might be relevant in the improved tests that will be addressed by the KLOE-2 experiment[13] in the minimal and nonminimal SME framework.

Acknowledgments

I would like to thank Alan Kostelecký and Ralf Lehnert for the invitation to CPT'19, and Ágnes Roberts for the pleasant stay in Bloomington; very interesting discussions with them on the subject are also acknowledged.

References

1. A. Di Domenico, ed., *Handbook on Neutral Kaon Interferometry at a ϕ Factory*, Frascati Physics Series **43**, INFN-LNF, Frascati, 2007.

2. J. Bernabeu, N. Mavromatos, and J. Papavassiliou, Phys. Rev. Lett. **92**, 131601 (2004).

3. F. Ambrosino *et al.*, Phys. Lett. B **642**, 315 (2006).

4. A. Di Domenico *et al.*, Found. Phys. **40**, 852 (2010).

5. A. Anastasi *et al.*, KLOE-2 Collaboration, J. High Energy Phys. **09**, 139 (2018).

6. J. Bernabeu, A. Di Domenico, and P. Villanueva, J. High Energy Phys. **10**, 139 (2015).

7. *Data Tables for Lorentz and CPT Violation*, V.A. Kostelecký and N. Russell, 2019 edition, arXiv:0801.0287v12.

8. V.A. Kostelecký, Phys. Rev. Lett. **80**, 1818 (1998); Phys. Rev. D **61**, 016002 (1999); Phys. Rev. D **64**, 076001 (2001).

9. KLOE Collaboration, A. Di Domenico, in V.A. Kostelecký, ed., *CPT and Lorentz Symmetry IV*, World Scientific, Singapore, 2008.

10. D. Babusci *et al.*, Phys. Lett. B **730**, 89 (2014).

11. A. Roberts, Phys. Rev. D **96**, 116015 (2017).

12. K. Schubert, private communication.

13. G. Amelino-Camelia *et al.*, Eur. Phys. J. C **68** 619 (2010).

Nonminimal Lorentz Violation in Linearized Gravity

Matthew Mewes

Physics Department, California Polytechnic State University
San Luis Obispo, CA 93407, USA

This contribution to the CPT'19 meeting provides a brief overview of recent theoretical studies of nonminimal Lorentz violation in linearized gravity. Signatures in gravitational waves from coalescing compact binaries are discussed.

The Standard-Model Extension (SME) is a general framework for studies of arbitrary realistic violations of Lorentz and CPT invariance. The SME has provided a theoretical base for hundreds of searches for Lorentz and CPT violations in particles and in gravity.[1] The leading-order violations in the particle sectors of the SME were written down more than two decades ago,[2] followed by the leading-order violations in gravity.[3] These violations modify the Standard Model of particle physics and General Relativity. Together they give the so-called minimal Standard-Model Extension (mSME).

A Lorentz-violating term in the SME action takes the form of a conventional tensor operator contracted with a tensor coefficient for Lorentz violation:

$$\delta S = \int d^4x \ (\text{coefficient tensor}) \cdot (\text{tensor operator}). \qquad (1)$$

The tensor coefficients for Lorentz violation act as Lorentz-violating background fields. Each violation can be classified according to the mass dimension d of the conventional operator in natural units with $\hbar = c = 1$. The mSME contains the violations of renormalizable dimensions $d = 3, 4$. Nonminimal violations are those with $d \geq 5$. Nonminimal extensions have been constructed for a number of sectors of the SME, including gauge-invariant electromagnetism,[4] neutrinos,[5] free Dirac fermions,[6] quantum electrodynamics,[7] General Relativity,[8,9] and linearized gravity.[9–14]

The extension for linearized gravity includes all possible modifications to the usual linearized Einstein–Hilbert action that are quadratic in the metric fluctuation $h_{\mu\nu} = g_{\mu\nu} - \eta_{\mu\nu}$. Each unconventional term takes the form[12]

$$\delta S = \int d^4 x \, \tfrac{1}{4} \mathcal{K}^{(d)\mu\nu\rho\sigma\alpha_1\ldots\alpha_{d-2}} h_{\mu\nu} \partial_{\alpha_1} \ldots \partial_{\alpha_{d-2}} h_{\rho\sigma}, \qquad (2)$$

where $\mathcal{K}^{(d)\mu\nu\rho\sigma\alpha_1\ldots\alpha_{d-2}}$ are the coefficients for Lorentz violation. The coefficient tensors can be split into irreducible pieces with unique symmetries, giving fourteen different classes of Lorentz violation.[13] Three of these classes yield modifications that are invariant under the usual gauge transformation $h_{\mu\nu} \to h_{\mu\nu} + \partial_{(\mu} \xi_{\nu)}$. Restricting attention to the gauge-invariant violations, the Lorentz-violating parts can be written as[12]

$$S_{\mathrm{LV}} = \int d^4 x \, \tfrac{1}{4} h_{\mu\nu} (\widehat{s}^{\mu\rho\nu\sigma} + \widehat{q}^{\mu\rho\nu\sigma} + \widehat{k}^{\mu\nu\rho\sigma}) h_{\rho\sigma}, \qquad (3)$$

where the three operators

$$
\begin{aligned}
\widehat{s}^{\mu\rho\nu\sigma} &= \sum s^{(d)\mu\rho\alpha_1\nu\sigma\alpha_2\ldots\alpha_{d-2}} \partial_{\alpha_1} \ldots \partial_{\alpha_{d-2}}, \\
\widehat{q}^{\mu\rho\nu\sigma} &= \sum q^{(d)\mu\rho\alpha_1\nu\alpha_2\sigma\alpha_3\ldots\alpha_{d-2}} \partial_{\alpha_1} \ldots \partial_{\alpha_{d-2}}, \\
\widehat{k}^{\mu\nu\rho\sigma} &= \sum k^{(d)\mu\alpha_1\nu\alpha_2\rho\alpha_3\sigma\alpha_4\ldots\alpha_{d-2}} \partial_{\alpha_1} \ldots \partial_{\alpha_{d-2}}
\end{aligned} \qquad (4)
$$

contain the three types of gauge-invariant violations. The s- and k-type violations are CPT even, while q-type violations break CPT invariance. The sums in Eq. (4) are over even $d \geq 4$ for s-type violations, odd $d \geq 5$ for q-type, and even $d \geq 6$ for k-type.

The gauge-invariant limit provides a simple framework for studies of Lorentz violation in gravity, including studies of short-range gravity,[9,10] gravitational Čerenkov radiation,[11] and gravitational waves.[12–14] For gravitational waves, the violations give a modified phase velocity of the form[12]

$$v = 1 - \varsigma^0 \pm \sqrt{|\varsigma_{(+4)}|^2 + |\varsigma_{(0)}|^2}. \qquad (5)$$

The effects of Lorentz violation are controlled by the frequency- and direction-dependent functions

$$
\begin{aligned}
\varsigma^0 &= \sum_{djm} \omega^{d-4}{}_0 Y_{jm}(-\hat{v}) \, k^{(d)}_{(I)jm}, \\
\varsigma_{(\pm4)} &= \sum_{djm} \omega^{d-4}{}_{\pm4} Y_{jm}(-\hat{v}) \, (k^{(d)}_{(E)jm} \pm i k^{(d)}_{(B)jm}), \\
\varsigma_{(0)} &= \sum_{djm} \omega^{d-4}{}_0 Y_{jm}(-\hat{v}) \, k^{(d)}_{(V)jm},
\end{aligned} \qquad (6)
$$

where ω is the angular frequency. Spin-weighted spherical harmonics ${}_s Y_{jm}$ are used to characterize the dependence on the direction of propagation $\hat{v}$.

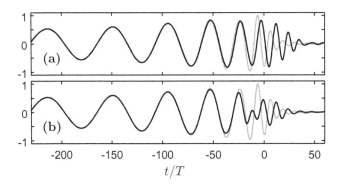

Fig. 1: Simulated noise-free detector strain signals from the merger of two equal-mass black holes for (a) nonbirefringent dispersion with $\varsigma^{(6)0} = 20T^3/\tau$ and (b) birefringence with $\varsigma^{(5)}_{(0)} = 10T^2/\tau$. The effective propagation time τ accounts for the redshift of the frequency during propagation. We also define the characteristic merger time scale $T = G_N(1+z)M$ in terms of Newton's constant G_N and the merger's redshift z and total mass M. The plots show the Lorentz-violating cases in black and the Lorentz-invariant limit in gray. [14]

The spherical coefficients for Lorentz violation $k^{(d)}_{(I)jm}$, $k^{(d)}_{(V)jm}$, $k^{(d)}_{(E)jm}$, and $k^{(d)}_{(B)jm}$ are complicated linear combinations of the underlying tensor coefficients in Eq. (4).

The Lorentz violation associated with $k^{(d)}_{(I)jm}$ coefficients produce a frequency-dependent velocity, leading to direction-dependent dispersion in gravitational waves. An example of the effects of this type of violation in a binary merger is shown in the top plot of Fig. 1. In this example, $d = 6$ violations lead to a shift in phase velocity that is proportional ω^2. This particular shift causes the higher-frequency components of the wave to travel slower than lower-frequency parts. Early in the merger, when lower frequencies dominate, the effects of dispersion are insignificant. Higher frequencies dominate at later times, where the signal experiences a delayed arrival, deforming the tail end of the waveform.

The violations associated with the $k^{(d)}_{(V)jm}$, $k^{(d)}_{(E)jm}$, and $k^{(d)}_{(B)jm}$ coefficients yield two distinct propagating solutions. The two solutions have different polarizations that are determined by the $\varsigma_{(+4)}$, $\varsigma_{(-4)}$, and $\varsigma_{(0)}$ combinations. [14] Each solution propagates at a different speed, corresponding to the two signs in Eq. (5). This gives rise to birefringence (in addition to dispersion). A general wave is a superposition of the two solutions, which results in a net polarization that evolves as the wave propagates, yielding a key signature of birefringence. The effects depend on frequency, so each frequency experiences a different change in polarization.

An example of the effects of birefringence are illustrated in the bottom plot of Fig. 1. In this example, $d = 5$ birefringent Lorentz violations produce a simple rotation of the polarization of the wave. The rotation angle grows with ω, so higher frequencies experience a greater change. The example assumes that the gravitational wave is linearly polarized and that the arms of the detector are aligned so that the strain signal is maximized in the Lorentz-invariant limit. At later stages in the merger, the higher frequencies produce a greater rotation of the polarization. This affects the relative alignment of the arms of the detector and the wave's polarization, decreasing the response of the detector and distorting the signal.

Acknowledgments

This work was supported in part by the US National Science Foundation under grant No. 1819412.

References

1. V.A. Kostelecký and N. Russell, *Data Tables for Lorentz and CPT Violation*, 2019 edition, arXiv:0801.0287v12.
2. V.A. Kostelecký and R. Potting, Phys. Rev. D **51**, 3923 (1995); D. Colladay and V.A. Kostelecký, Phys. Rev. D **55**, 6760 (1997); Phys. Rev. D **58**, 116002 (1998).
3. V.A. Kostelecký, Phys. Rev. D **69**, 105009 (2004); Q.G. Bailey and V.A. Kostelecký, Phys. Rev. D **74**, 045001 (2006); V.A. Kostelecký and J. Tasson, Phys. Rev. D **83**, 016013 (2011).
4. V.A. Kostelecký and M. Mewes, Phys. Rev. Lett. **99**, 011601 (2007); Ap. J. **689**, L1 (2008); Phys. Rev. D **80**, 015020 (2009); Phys. Rev. Lett. **110**, 201601 (2013).
5. V.A. Kostelecký and M. Mewes, Phys. Rev. D **85**, 096005 (2012).
6. V.A. Kostelecký and M. Mewes, Phys. Rev. D **88**, 096006 (2013).
7. Y. Ding and V.A. Kostelecký, Phys. Rev. D **94**, 056008 (2016); V.A. Kostelecký and Z. Li, Phys. Rev. D **99**, 056016 (2019).
8. Q.G. Bailey, Phys. Rev. D **94**, 065029 (2016).
9. Q.G. Bailey, V.A. Kostelecký, and R. Xu, Phys. Rev. D **91**, 022006 (2015).
10. V.A. Kostelecký and M. Mewes, Phys. Lett. B **766**, 137 (2017).
11. V.A. Kostelecký and J.D. Tasson, Phys. Lett. B **749**, 551 (2015).
12. V.A. Kostelecký and M. Mewes, Phys. Lett. B **757**, 510 (2016).
13. V.A. Kostelecký and M. Mewes, Phys. Lett. B **779**, 136 (2018).
14. M. Mewes, Phys. Rev. D **99**, 104062 (2019).

Status and Prospects for CPT Tests with the ALPHA Experiment

T. Friesen

Department of Physics and Astronomy, University of Calgary,
Calgary, Alberta T2N1N4, Canada

On behalf of the ALPHA Collaboration

A primary goal of the ALPHA experiment at CERN is to perform precise tests of CPT symmetry. Here, we report on the significant progress made in recent years on antihydrogen spectroscopy and the outlook for the future.

1. Introduction

The ALPHA collaboration is focused on precision measurements with antihydrogen ($\bar{\text{H}}$) to test fundamental symmetries between matter and antimatter. The H–$\bar{\text{H}}$ system is natural for such tests because of its relative simplicity and its key role in the development of modern physics. Spectroscopy of H has a long and successful history: the 1S–2S transition and the ground-state hyperfine splitting (GSHFS) have been measured at the levels of 10^{-15} and 10^{-12}, respectively.[1,2] Since CPT symmetry implies the equality of the H and $\bar{\text{H}}$ spectra, reaching similar sensitivities in $\bar{\text{H}}$ represents an excellent CPT test in a purely atomic–anti-atomic system. Signals for CPT and Lorentz violation can be described using the Standard-Model Extension (SME).[3] The SME is a realistic effective field theory built from General Relativity, the Standard Model, and all possible Lorentz-violating operators.

2. Antihydrogen trapping and detection

The Antiproton Decelerator at CERN provides ALPHA with 3×10^7 antiprotons ($\bar{p}$) every ~ 100 s. We capture roughly 90,000 of these in a Penning–Malmberg trap, where we also load 3×10^6 e^+ from a Surko-type accumulator.[4] The $\bar{p}$ and e^+ are separately cooled, compressed, and then mixed to form roughly 50,000 $\bar{\text{H}}$ atoms. To confine $\bar{\text{H}}$, ALPHA employs a magnetic-minimum neutral-atom trap formed by two short solenoids for axial confinement and an octupole winding for transverse confinement. In addition, there are three central short solenoids used to flatten the axial $\vec{B}$ field near the trap's center to aid with spectroscopy. Of the 50,000 atoms formed, an average of only 10–20 $\bar{\text{H}}$ have a low enough kinetic energy to be trapped. $\bar{\text{H}}$ atoms from consecutive mixing cycles can be accumulated, and hundreds of $\bar{\text{H}}$ atoms can be loaded into the trap in this manner.[5]

The annihilation of unconfined $\bar{\text{H}}$ is detected by a surrounding three-layer silicon annihilation detector. The $\bar{p}$ annihilation on the Penning-trap electrodes will produce an average of three charged pions that register hits on each layer of silicon and allow the reconstruction of an annihilation vertex. The main background comes from cosmic rays that trigger the detector at a rate of $10 \pm 0.02\,\text{s}^{-1}$. Because the topology of the signal (annihilations) and background (cosmic rays) events are very different, they can be distinguished effectively by using machine-learning procedures.[6]

3. Antihydrogen spectroscopy

In the past several years, improved particle preparation techniques[5,7] have drastically increased $\bar{\text{H}}$ production and trapping rates at ALPHA, opening up numerous avenues for performing experiments on $\bar{\text{H}}$ including 1S–2S, hyperfine, and 1S–2P spectroscopy. Below is a brief description of these efforts.

ALPHA's primary spectroscopic goal has been 1S–2S spectroscopy because of the high precision achieved in H. The 1S–2S transition requires two simultaneous 243 nm photons and because we are interacting with small numbers of $\bar{\text{H}}$ (at most hundreds), relatively high power 243 nm radiation is needed. For these reasons, the ALPHA apparatus includes a cryogenic build-up cavity that allows the input power of 160 mW to be built up to $\sim 1\,\text{W}$ of circulating power. After two-counter propagating photons excite the atom to the 2S state, absorption of a third photon ionizes the atom leading to loss and annihilation of the $\bar{p}$ from the trap. In addition, atoms can couple from the 2S state to the 2P state and then be lost if they undergo a e^+ spin flip during decay back down to the 1S state. Both of these mechanisms lead to an annihilation signal indicating that an excitation to 2S occurred. In 2018, ALPHA published a measurement of the 1S–2S transition in $\bar{\text{H}}$ in a 1 T field to a precision of 5×10^3 Hz out of 2.5×10^{15} Hz.[10] This is consistent with CPT invariance at a relative precision of 2×10^{-12} (corresponding to an energy sensitivity of 2×10^{-20} GeV). This result has been used to put a constraint on CPT-violating SME coefficients, the first such constraint from $\bar{\text{H}}$ spectroscopy.[11]

Also of interest is the GSHFS in H, which is sensitive to different SME parameters than the 1S–2S transition and may potentially be even more sensitive to CPT-violating effects despite the lower relative precision of the measurement.[8] The ground state of $\bar{\text{H}}$ in a strong $\vec{B}$ field is split into two pairs of states. One pair, the high-field seekers $|a\rangle$ and $|b\rangle$ with their e^+ spins aligned with $\vec{B}$, is not trapped by the magnetic-minimum trap. The other pair, the low-field seekers $|c\rangle$ and $|d\rangle$ with their e^+ spins anti-aligned with $\vec{B}$, is trapped. To measure the GSHFS, we excite the two positron spin-resonance transitions $|c\rangle \to |b\rangle$ and $|d\rangle \to |a\rangle$ and measure their frequencies. If both measurements are performed at the same $\vec{B}$ field, we can find the GSHFS through $f_{\text{HFS}} = f_{da} - f_{cb}$. At a base field of $\sim 1\,\text{T}$, these transitions occur at $\sim 29\,\text{GHz}$. When an $\bar{\text{H}}$ undergoes such a transition, it is put into a high-field seeking state and will quickly annihilate on the surrounding apparatus walls. These annihilations are registered by the detector and used as the signal that a transition occurred.

In 2017, ALPHA published the first measurement[9] of the GSHFS in $\bar{\text{H}}$ with a precision of four parts in 10^4. Since that time, the rate at which trapped $\bar{\text{H}}$ can be produced and accumulated has increased considerably. Improvements have also been made to our ability to control, stabilize, and measure the trapping $\vec{B}$ fields. With more atoms and better field control this measurement can be improved significantly.

A third transition that has been a major focus for ALPHA is the 1S–2P transition, which could be used to laser cool $\bar{\text{H}}$ for improved spectroscopy and gravity measurements. To excite this transition, narrow line-width (roughly 65 MHz) pulsed (about 12 ns duration) laser light at 121.6 nm is generated by third-harmonic generation of 365 nm light in a high-pressure Kr/Ar gas cell. Each pulse has an energy of 0.53–0.63 nJ in the trapping region, and the pulse repetition rate is 10 Hz. After being excited to 2P, the atoms decay back to 1S within a few ns with a probability to undergo a e^+ spin flip and subsequently annihilate on the surrounding walls. In 2018, ALPHA published the results[12] of an experiment demonstrating the excitation of 1S–2P transitions. Based on a dataset consisting of 966 detected annihilations, ALPHA observed the 1S–2P transition in $\bar{\text{H}}$ and determined the transition frequency to a relative precision of 5×10^{-8}. With this demonstration, we are now in a position to attempt to laser cool $\bar{\text{H}}$. Simulations predict that in the geometry of the current apparatus, laser cooling to roughly 20 mK is possible.[13]

64

4. Outlook

After the major successes of recent years, ALPHA will continue to push
H̄ spectroscopy to new precisions, explore new transitions, and measure
at different $\vec{B}$ fields. A further transition of interest is the $\bar{p}$ spin flip
transition between the $|c\rangle$ and $|d\rangle$ states. This transition exhibits a broad
maximum near 0.65 T making it less sensitive to $\vec{B}$ fields, which is a major
source of uncertainty. Two-photon 1S–2S spectroscopy in the near term
can be improved by using larger waist size for the radiation in the optical
cavity to reduce transit-time broadening. Also, laser cooling of the atoms
down to ~ 20 mK would greatly narrow the measured linewidth. With the
demonstration of our ability to excite the 1S–2P transition, laser cooling
of H̄ is within reach. Finally, ALPHA is also building a new apparatus
to measure the gravitational free-fall of H̄. This new apparatus, known as
ALPHAg, is a vertical magnetic trap that will initially allow us to determine
if H̄ falls up or down upon release and ultimately aims to measure the
gravitational mass of H̄ at the 1% level.

Acknowledgments

This work was supported by: the European Research Council through its
Advanced Grant programme (JSH); CNPq, FAPERJ, RENAFAE (Brazil);
NSERC, NRC/TRIUMF, EHPDS/EHDRS (Canada); FNU (NICE Cen-
tre), Carlsberg Foundation (Denmark); ISF (Israel); STFC, EPSRC, Royal
Society, Leverhulme Trust (UK); DOE, NSF (USA); and VR (Sweden).

References

1. C.G. Parthey *et al.*, Phys. Rev. Lett. **107**, 203001 (2011).
2. P. Petit *et al.*, Metrologia **16**, 7 (1980).
3. V.A. Kostelecký and A.J. Vargas, Phys. Rev. D **92**, 056002 (2015).
4. T.J. Murphy and C.M. Surko, Phys. Rev. A **46**, 5696 (1992).
5. ALPHA Collaboration, M. Ahmadi *et al.*, Nat. Commun. **8**, 681 (2017).
6. ALPHA Collaboration, M. Ahmadi *et al.*, Nature **557**, 71 (2018); A. Hoecker *et al.*, arXiv:physics/0703039.
7. ALPHA Collaboration, M. Ahmadi *et al.*, Phys. Rev. Lett. **120**, 025001 (2018).
8. R. Bluhm *et al.*, Phys. Rev. Lett. **82**, 2254 (1999).
9. ALPHA Collaboration, M. Ahmadi *et al.*, Nature **548**, 66 (2017).
10. ALPHA Collaboration, M. Ahmadi *et al.*, Nature, **557**, 71 (2018).
11. V.A. Kostelecký and A.J. Vargas, Phys. Rev. D **98**, 036003 (2018).
12. ALPHA Collaboration, M. Ahmadi *et al.*, Nature **561**, 211 (2018).
13. P.H. Donnan *et al.*, J. Phys. B **46**, 025302 (2013).

Precision Tests of Lorentz Invariance and Fundamental Physics with Acoustic Phonons

M.E. Tobar, M. Goryachev, and E.N. Ivanov

ARC Centre of Excellence For Engineered Quantum Systems,
Department of Physics, University of Western Australia,
35 Stirling Highway, Crawley WA 6009, Australia

Quartz Bulk Acoustic Wave resonators and oscillators are amongst the best devices for frequency-control applications. For instance, at the University of Western Australia we have set up new experiments to test for Lorentz-invariance violations, to search for Dark Matter, to search for high-frequency gravitational-wave radiation, and implement macroscopic acoustic oscillators to undertake tests of quantum gravity. The Lorentz test uses stable room-temperature oscillators on a rotating platform, while the other techniques utilize cryogenic resonators cooled to low temperatures. At cryogenic temperatures, such resonators have been shown to have record acoustic Q factors of more than a billion near $4\,\mathrm{K}$ and above $100\,\mathrm{MHz}$.

1. Introduction

Recently, very high Q-factor quartz Bulk Acoustic Wave (BAW) resonators have been shown to have the highest $Q \cdot f$ products of any resonant mechanical system.[1,2] Correspondingly, they have attracted interest for quantum measurements and fundamental-physics tests.[3–6] In this work, we summarize our use of such high-Q resonators and oscillators at room temperature and low temperatures to test fundamental physics.

2. Fundamental-physics tests

Currently, we have experiments set up and running based on quartz BAW resonators to conduct some new tests of fundamental physics; these include:

(1) *Rotating experiment to test Lorentz symmetry.*[5,7] This is a second-generation experiment, which technically operates with two orders of magnitude better sensitivity than the previous version. It has been running for nearly a year, and after the year finishes, we will be able to set new limits on the corresponding Lorentz-violating coefficients.

(2) *Cryogenic BAW resonator coupled to a SQUID amplifier.* These experiments are limited by Nyquist thermal or quantum fluctuations.[4] Such a system is sensitive to gravitational waves[8] and dark matter[9] and can be implemented for testing quantum gravity.[10] Currently, such an experiment has been under operation for several months and will be used to search for new physic in the near future.

3. Rotating experiment to test Lorentz invariance

In the past, we have implemented oscillators based on quartz BAW resonators for a high-precision test of Lorentz invariance.[5] The first experiment utilized oscillators inferior to the best available, and recently we have began a second-generation Lorentz-invariance experiment based on the best available room-temperature oscillators with technological improvements in the rotation set up.[7] The experiment uses room-temperature oscillators with state-of-the-art phase noise that are continuously compared on a platform that rotates at a rate of the order of a cycle per second. The new system includes improvements in noise-measurement techniques, data acquisition, and data processing. Preliminary results from this experiment indicate that Standard-Model Extension coefficients in the matter sector can be measured at a precision of order $10^{-16}\,$GeV after taking a year's worth of data. This is equivalent to an improvement of two orders of magnitude over the prior acoustic-phonon-sector experiment.[5] Currently, the experiment has been running for nearly a year, and soon we will put limits on Standard-Model Extension coefficients in the nonminimal matter sector.

On might expect cryogenic versions of the oscillators to yield better results due to the extremely high Q factors. However, we have realized a Pound-stabilized cryogenic oscillator from one such mode near 116 MHz. Despite the extremely high Q factors, we find that the power–frequency sensitivity is increased by a factor of 1000 at cryogenic temperatures, which limits the purity of the oscillator signal at low temperatures. This problem would need to be dealt with in the future for a cryogenic system to surpass a room-temperature one.

4. Macroscopic acoustic resonators to test fundamental physics at low temperatures

Rather than oscillators as discussed previously, macroscopic acoustic resonators with a low noise readout also make good test beds for many fundamental-physics tests. The original high-precision measurement of this type was the resonant-mass gravitational-wave (GW) detector.[11] Since then, the quantum-information discipline has further pushed this technology, and many other ideas and systems have been proposed. Such systems can now operate at the quantum limit and surpass this limit in sensitivity. In particular, we have built such experiments based on high-Q modes in sapphire[12] and quartz,[4] and we are using such setups to test also quantum gravity.[10]

In the context of searches for high-frequency GWs, we harness phonons by implementing phonon-trapping BAW resonator technology.[8] Extremely sensitive experiments are possible due to the extremely high Q factors at cryogenic temperatures, $Q \approx 10^{10}$, for frequencies ranging from 5 MHz to nearly a GHz, which is beyond the capability of any other technology. We have shown that a new, highly sensitive GW search is possible over this frequency range, where no prior search has been undertaken. The experiment consists of a BAW resonator coupled to a SQUID amplifier, with two independent cryogen-free cooled experiments initially at 4 K.[4] Two independent systems are needed to look for the signal correlations necessary to confirm the existence of GWs similar to the way the two LIGO detectors operate. Furthermore, we propose in the future to have two BAW resonators oriented orthogonally in each experiment, so we can reject spurious signals originating from within each cryogen-free system. This will reduce the background and improve detection confidence.

This is a low-cost experiment with a high potential gain and no risk. A first detection would be of major scientific importance and a null result will still attract considerable interest, as there are a number of theoretical predictions for astrophysical and cosmological objects at these frequencies that this experiment can either verify or rule out.[8] The aim is to perform the first search in a 4 K environment with a sensitivity of $10^{-21}/\sqrt{\mathrm{Hz}}$ per mode (in the future, a more sensitive mK experiment is possible).

5. Search for scalar dark matter

The same set up to search for GWs is also sensitive to scalar dark matter, as shown by Arvanitaki et al., in "Sound of Dark Matter: Searching for Light Scalars with Resonant-Mass Detectors."[9] Scalar fields called moduli determine the fine-structure constant and electron mass in string theory. They show that our quartz GW experiment can put limits on dark matter if it takes on the form of such a light modulus and will oscillate with a frequency equal to its mass and amplitude determined by the local dark-matter density. This translates into an oscillation of the size of a solid that can be observed by resonant-mass GW antennas, and hence the phonon modes in quartz BAW resonators. The data analysis required to put limits on such oscillations is very similar to searching for a coherent source of GWs.

Acknowledgments

This work was funded by the Australian Research Council under grant No. CE170100009.

References

1. S. Galliou, M. Goryachev, R. Bourquin, P. Abbé, J.P. Aubry, and M.E. Tobar, Sci. Rep. **3**, 2132 (2013).
2. I. Pikovski, M.R. Vanner, M. Aspelmeyer, M.S. Kim, and C. Brukner, Nat. Phys. **8**, 393 (2012).
3. M. Goryachev, D. Creedon, S. Galliou, and M.E. Tobar, Phys. Rev. Lett. **111**, 085502 (2013).
4. M. Goryachev, E.N. Ivanov, F. van Kann, S. Galliou, and M.E. Tobar, Appl. Phys. Lett. **105**, 153505 (2014).
5. A. Lo, P. Haslinger, E. Mizrachi, L. Anderegg, H. Mller, M. Hohensee, M. Goryachev, and M.E. Tobar, Phys. Rev. X **6**, 011018 (2016).
6. M. Goryachev and M.E. Tobar, New J. Phys. **16**, 083007 (2014).
7. M. Goryachev, Z. Kuang, E.N. Ivanov, P. Haslinger, H. Müller, and M.E. Tobar, IEEE Trans. Ultrason., Ferroelectr., Freq. Control **65**, 991 (2018).
8. M. Goryachev and M.E. Tobar, Phys. Rev. D **90**, 102005 (2014).
9. A. Arvanitaki, S. Dimopoulos, and K. Van Tilburg, Phys. Rev. Lett. **116**, 031102 (2016).
10. P.A. Bushev, J. Bourhill, M. Goryachev, N. Kukharchyk, E.N Ivanov, S. Galliou, M.E. Tobar, and S. Danilishin, arXiv:1903.03346.
11. D.G. Blair, E.N. Ivanov, M.E. Tobar, P.J. Turner, F. van Kann, and I.S. Heng, Phys. Rev. Lett. **74**, 1908 (1995).
12. J. Bourhill, E.N. Ivanov, and M.E. Tobar, Phys. Rev. A **92**, 023817 (2015).

Recent Developments in Spacetime-Symmetry Tests in Gravity

Quentin G. Bailey

Department of Physics and Astronomy, Embry-Riddle Aeronautical University,
Prescott, AZ 86301, USA

We summarize theoretical and experimental work on tests of CPT and local Lorentz symmetry in gravity. Recent developments include extending the
effective-field-theory framework into the nonlinear regime of gravity.

1. Introduction

Motivated by potentially detectable but minuscule signatures from Planck-
scale or other new physics, there has been a substantial increase in tests of
spacetime symmetries in gravity in recent years. [1,2] Some novel hypothetical
effects that break local Lorentz symmetry and CPT symmetry in gravitational experiments as well as solar-system and astrophysical observations
have been studied in recent works. [3] Much of this work uses the effective-
field-theory framework called the Standard-Model Extension (SME), which
includes gravitational couplings. [4,5] In other cases, the parameters in specific hypothetical models of Lorentz violation in gravity have been tested. [6]

2. Framework

The general framework of the SME in the pure-gravity sector can be realized as the Einstein–Hilbert action plus a series of terms formed from
indexed coefficients, explicit or dynamical, contracted with increasing powers of curvature and torsion. Each term in this series maintains observer
invariance of physics, while breaking "particle" invariance with respect to
local Lorentz symmetry and diffeomorphism symmetry. [5]

One interesting and practical subset of the SME is a general description
of CPT and Lorentz violation that is provided by an expansion valid for
linearized gravity ($g_{\mu\nu} = \eta_{\mu\nu} + h_{\mu\nu}$). For instance, in this approximation
the Lagrange density for General Relativity (GR) plus the mass dimension
four and five operators controlling local Lorentz and CPT violation are

given by[7-9]

$$\mathcal{L} = -\frac{1}{4\kappa}(h^{\mu\nu}G_{\mu\nu} - \overline{s}^{\mu\kappa}h^{\nu\lambda}\mathcal{G}_{\mu\nu\kappa\lambda} + \frac{1}{4}h_{\mu\nu}(q^{(5)})^{\mu\rho\alpha\nu\beta\sigma\gamma}\partial_\beta R_{\rho\alpha\sigma\gamma} + ...), \quad (1)$$

where $\kappa = 8\pi G_N$, and the double-dual curvature $\mathcal{G}$ and the Riemann curvature $R_{\rho\alpha\sigma\gamma}$ are linearized in $h_{\mu\nu}$. This Lagrange density maintains linearized diffeomorphism invariance, though generalizations exist,[10] and $\overline{s}_{\mu\nu}$ and $(q^{(5)})^{\mu\rho\alpha\nu\beta\sigma\gamma}$ are the coefficients controlling the degree of symmetry breaking (they are zero in GR).

3. Experiment and observation

The mass dimension $d = 4$ Lagrange density, the minimal gravity SME, has now been studied in a plethora of tests. The best controlled and simultaneous parameter-fitting limits come from lunar laser ranging,[11] and other laboratory experiments such as gravimetry.[12] These place limits on the $\overline{s}_{\mu\nu}$ coefficients at the level of approximately $10^{-7} - 10^{-8}$ on the three $\overline{s}_{TJ}$ and $10^{-10} - 10^{-11}$ on five of the $\overline{s}_{JK}$ coefficients. Stronger limits can be countenanced from distant cosmic rays,[13] and one combination of coefficients is bounded at 10^{-15} by the multimessenger neutron-star inspiral event in 2017.[14] Other searches for these coefficients include ones with pulsars.[15]

For the mass dimension $d = 5$ coefficients in (1) that break CPT symmetry, the post-Newtonian phenomenology includes a velocity-dependent inverse cubic force. This leads to an extra term in the relative acceleration of two bodies given by[16]

$$\delta a^j = \frac{G_N M v^k}{r^3}\left(15n^l n^m n^n n_{[j}K_{k]lmn}\right.$$
$$\left. + 9n^l n^m K_{[jk]lm} - 9n_{[j}K_{k]llm}n^m - 3K_{[jk]ll}\right), \quad (2)$$

where K_{jklm} are linear combinations of the coefficients q in the Lagrange density (1), $\vec{r}$ is the separation between the bodies and $\hat{n} = \vec{r}/r$.

Measurements of the mass dimension $d = 5$ coefficients in (2) are currently scarce. There is one constraint on a combination of dimension five and six coefficients from Ref. 9 in searches for dispersion of gravitational waves from distant sources and an analysis with multiple gravitational-wave events is underway.[17] Disentangled constraints on the K_{jklm} coefficients from analyses of pulsar observations exist at the level of 10^6 m.[18] This leaves room for potentially large, "countershaded" symmetry breaking to exist in nature.[19] Higher-order terms in the series, at mass dimension $d = 6$ and beyond, have been constrained in short-range gravity tests.[20]

4. Extension to the nonlinear regime

While the general form for linearized gravity has been explored, only several works have explored the general SME framework beyond linearized gravity.[21] One approach is to extend the general Lagrange density for linearized gravity (which is of quadratic order in the metric fluctuations) to include terms of cubic and higher order. If we adopt the point of view of spontaneous symmetry breaking (SSB), one must consider the dynamics of the coefficients for Lorentz violation. Considering the case of a symmetric 2-tensor $s_{\mu\nu}$ being the Lorentz-breaking field, it is expanded in the SSB scenario as $s_{\mu\nu} = \bar{s}_{\mu\nu} + \tilde{s}_{\mu\nu}$, where $\bar{s}_{\mu\nu}$ are the vacuum expectation values and $\tilde{s}_{\mu\nu}$ are the fluctuations. The Lagrange density is a series $\mathcal{L} = \mathcal{L}^{(2)} + \mathcal{L}^{(3)} + ...$ where (2) and (3) indicate the order in fluctuations $h_{\mu\nu}$ or $\tilde{s}_{\mu\nu}$. A general conservation law,[22] contained in Eq. (9) of Ref. 23, can be used to constrain the terms in the series. In the example of $s_{\mu\nu}$, it takes the form

$$\partial_\beta \left(\frac{\delta\mathcal{L}}{\delta h_{\gamma\beta}} \right) + \Gamma^\gamma{}_{\alpha\beta} \left(\frac{\delta\mathcal{L}}{\delta h_{\alpha\beta}} \right) + g^{\delta\gamma} s_{\delta\alpha} \partial_\beta \left(\frac{\delta\mathcal{L}}{\delta \tilde{s}_{\alpha\beta}} \right) + g^{\delta\gamma} \tilde{\Gamma}_{\delta\alpha\beta} \frac{\delta\mathcal{L}}{\delta h_{\alpha\beta}} = 0, \quad (3)$$

where $\tilde{\Gamma}_{\delta\alpha\beta} = (\partial_\alpha \tilde{s}_{\beta\delta} + \partial_\beta \tilde{s}_{\alpha\delta} - \partial_\delta \tilde{s}_{\alpha\beta})/2$. This equation holds "off-shell," assuming the action obtained from $\mathcal{L}$ is diffeomorphism invariant.

In the case of the minimal SME with just $s_{\mu\nu}$, the Lagrange density is constructed from all possible contractions of generic terms of the quadratic form $\bar{s}_{\alpha\beta} h_{\gamma\delta} \partial_\epsilon \partial_\zeta h_{\eta\theta}, \bar{s}_{\alpha\beta} \partial_{\gamma\alpha} \partial_\delta \tilde{s}_{\epsilon\zeta}, \bar{s}_{\alpha\beta} \partial_\gamma \partial_\delta h_{\epsilon\zeta}, ...,$ the cubic form $\bar{s}_{\alpha\beta} h_{\gamma\delta} h_{\epsilon\zeta} \partial_\eta \partial_\theta h_{\kappa\lambda}, \bar{s}_{\alpha\beta} h_{\gamma\delta} \partial_\epsilon h_{\zeta\eta} \partial_\theta h_{\kappa\lambda}, h_{\alpha\beta} \tilde{s}_{\gamma\delta} \partial_\epsilon \partial_\zeta \tilde{s}_{\theta\kappa}, ...,$ and potential terms. The sum of all such terms, each with an arbitrary parameter, is inserted into (3) and the resulting linear equations for the parameters are solved. What remains, up to total-derivative terms in the action, are a set of independently diffeomorphism-invariant terms. As an example of such a term produced by this expansion, we find to cubic order

$$\mathcal{L} \supset \bar{s}_{\alpha\beta} \tilde{s}^{\alpha\beta} R^{(1)} + \tfrac{1}{2} \tilde{s}_{\alpha\beta} \tilde{s}^{\alpha\beta} R^{(1)} - 2h^{\alpha\beta} \bar{s}_\alpha{}^\gamma \tilde{s}_{\beta\gamma} R^{(1)}$$

$$+ \bar{s}_{\alpha\beta} \tilde{s}^{\alpha\beta} (\Gamma_{\gamma\delta\epsilon} \Gamma^{\gamma\delta\epsilon} - \Gamma^{\gamma\delta}{}_\delta \Gamma_{\gamma\epsilon}{}^\epsilon + \tfrac{1}{2} h^\gamma{}_\gamma R^{(1)} - 2h^{\gamma\delta} R^{(1)}_{\gamma\delta}), \quad (4)$$

where the (1) superscript indicates linear order in $h_{\mu\nu}$ and the connection coefficients are at linear order. Note that this construction generally includes dynamical terms for the fluctuations and so does not assume "decoupling."[24]

The construction including all such terms allows exploration of the regime in gravity where nonlinearities need to be considered.[25] This includes higher-order post-Newtonian gravity in weak-field systems and developing a multipole expansion for gravitational waves affected by Lorentz violation.

Acknowledgments

This work was supported by the US National Science Foundation under grant No. 1806871 and Embry-Riddle Aeronautical University's FIRST grant program.

References

1. V.A. Kostelecký and S. Samuel, Phys. Rev. D **39**, 683 (1989).
2. *Data Tables for Lorentz and CPT Violation,* V.A. Kostelecký and N. Russell, 2019 edition, arXiv:0801.0287v12.
3. C.M. Will, Living Rev. Rel. **17**, 4 (2014); J.D. Tasson, Rept. Prog. Phys. **77**, 062901 (2014); A. Hees *et al.*, Universe **2**, 30 (2016).
4. D. Colladay and V.A. Kostelecký, Phys. Rev. D **55**, 6760 (1997); Phys. Rev. D **58**, 116002 (1998).
5. V.A. Kostelecký, Phys. Rev. D **69**, 105009 (2004).
6. N. Yunes *et al.*, Phys. Rev. D **94**, 084002 (2016).
7. Q.G. Bailey and V.A. Kostelecký, Phys. Rev. D **74**, 045001 (2006).
8. Q.G. Bailey *et al.*, Phys. Rev. D **91**, 022006 (2015).
9. V.A. Kostelecký and M. Mewes, Phys. Lett. B **757**, 510 (2016).
10. V.A. Kostelecký and M. Mewes, Phys. Lett. B **779**, 136 (2018).
11. A. Bourgoin *et al.*, Phys. Rev. Lett. **117**, 24130 (2016).
12. H. Müller *et al.*, Phys. Rev. Lett. **100**, 031101 (2008); N.A. Flowers *et al.*, Phys. Rev. Lett. **119**, 201101 (2017); C.-G. Shao *et al.*, Phys. Rev. D **97**, 024019 (2018).
13. V.A. Kostelecký and J.D. Tasson, Phys. Lett. B **749**, 551 (2015).
14. B.P. Abbott *et al.*, Astrophys. J. **848**, L13 (2017).
15. L. Shao, Phys. Rev. Lett. **112**, 111103 (2014).
16. Q.G. Bailey and D. Havert, Phys. Rev. D **96**, 064035 (2017).
17. M. Mewes, Phys. Rev. D **10**, 104062 (2019); K. O'Neal-Ault, these proceedings.
18. L. Shao and Q.G. Bailey, Phys. Rev. D **98**, 084049 (2018).
19. V.A. Kostelecký and J. Tasson, Phys. Rev. Lett. **102**, 010402 (2009).
20. C.-G. Shao *et al.*, Phys. Rev. Lett. **117**, 071102 (2016); Phys. Rev. Lett. **122**, 011102 (2019).
21. Y. Bonder, Phys. Rev. D **91**, 125002 (2015); Y. Bonder and G. León, Phys. Rev. D **96**, 044036 (2017); N.A. Nilsson *et al.*, these proceedings.
22. R. Bluhm, Phys. Rev. D **91**, 065034 (2015).
23. R. Bluhm and A. Sehic, Phys. Rev. D **94**, 104034 (2016).
24. M. Seifert, Phys. Rev. D **79**,124012 (2009); Symmetry **10**, 490 (2018).
25. Q.G. Bailey, to appear.

GINGER

Angela D.V. Di Virgilio

INFN Sez. di Pisa, Largo B. Pontecorvo 3, Pisa, Italy

On behalf of the GINGER Collaboration[*]

GINGER (Gyroscopes IN GEneral Relativity), based on an array of large-dimension ring-laser gyroscopes, is aiming at measuring in a ground laboratory the gravito-electric and gravito-magnetic effects (also known as the de Sitter and the Lense–Thirrings effect), predicted by General Relativity. The sensitivity depends on the size of the ring-laser-gyroscope cavities and the cavity losses, considering the present sensitivity, and assuming total losses of 6 ppm, with a 40-m perimeter and one day of integration time, a sensitivity at the order of frad/s is attainable. The construction of GINGER is at present under discussion.

The Sagnac effect has been discovered by Georges Sagnac more than 100 years ago, and states that the difference of time of flight of two light beams counter-propagating along a closed path is proportional to the angular rotation rate of the frame. In 2014 the Institute of France has organized a symposium to celebrate the centennial of this discovery; the symposium book provides a wide picture of this effect and its scientific applications.[1] In general, the Sagnac signal is generated by the time-of-flight difference between the two counter-propagating waves along the closed path, and is related to any non-reciprocity between the two propagation directions. The most general formula is:

$$\Delta\phi = \tfrac{8\pi A\Omega}{\lambda c}\cos\theta, \tag{1}$$

[*]C. Altucci,[2] A. Basti,[3] N. Beverini,[3] F. Bosi,[1] G. Carelli,[3] D. Ciampini,[3] F. Fuso,[3] U. Giacomelli,[1] E. Maccioni,[1,3] A. Ortolan,[4] A. Porzio,[5] A. Simonelli,[1] F. Stefani,[3] G. Terreni,[1] R. Velotta[1]

[1]INFN Sez. di Pisa, Largo B. Pontecorvo 3, Pisa, Italy
[2]Dept. of Physics, Univ. "Federico II" and INFN, Napoli, via Cintia, 80126 Napoli, Italy
[3]Dept. of Physics of the Univ. of Pisa, Largo B. Pontecorvo 3, Pisa, Italy
[4]INFN - National Lab. of Legnaro, viale dell'Università 2, 35020 Legnaro, Italy
[5]CNR - SPIN and INFN, Napoli, via Cintia, 80126 Napoli, Italy

74

where $\Delta\phi$ is the difference in phase of the two output waves, A is the enclosed area, λ the wavelength, θ the angle between the area vector of the ring and the rotational axis of the angular velocity $\vec{\Omega}$, and c the speed of light. In this case, the scale factor S_0 is $S_0 = \frac{8\pi A}{\lambda c}$.

Thus far, several different probes have been utilized: light, atoms, and superfluid helium.[1] Here, we limit the discussion to light as a probe, which, for high-sensitivity applications, has the obvious advantage of being insensitive to gravity variations. The closed path can be an optical-fiber coil or a ring Fabry–Perot cavity.[a] The devices based on resonant Fabry–Perot ring cavities have very interesting properties. They can be passive or active. Passive is when the resonant cavity is interrogated injecting light from the outside (passive ring cavity, PRC); active is when it contains an active medium and the device is a laser emitting two counter-propagating modes. In this case, it is called Ring Laser Gyro (RLG) or active ring cavity (ARC).[2]

The output of an RLG is the beat frequency f_s of the two modes, which is related to $\vec{\Omega}$:

$$f_s = S\Omega\cos\theta, \quad S = 4\frac{A}{\lambda L} \tag{2}$$

where L is the perimeter and S the geometrical scale factor. This is advantageous since the frequency measurement is extremely accurate and provides a large dynamic range. Moreover the scale factor S can be more efficiently stabilized, since it is proportional to A/L; this ratio can change with time. Building the apparatus in order to have this ratio close to a saddle point, the requirement on the long-time stability of the geometry can be relaxed.[b]

At present, Earth-based RLGs with perimeters of several meters are by orders of magnitude the most sensitive, both for short-time and long-time measurements; PRC are presently approaching the nrad/s sensitivity for a one-second integration time, atom-based gyros have recently reached 0.3 nrad/s levels for three hours of integration time, and the best RLGs[2] have sensitivities of 10 prad/s for a one-second integration time and can integrate for one day reaching 0.1 prad/s.[2] They have applications for geodesy, geophysics, and for tests of General Relativity (GR).[1] One example is GINGERINO, a 14-m perimeter RLG based on a simple mechanical structure that can be easily built and oriented at will, which runs unat-

---

[a]These two concept are widely used to develop inertial navigation gyroscopes.
[b]Very low frequency signals are at the base of this kind of research. As a consequence, long-time stability of the apparatus is necessary.

tended and uncontrolled for months with a duty cycle higher than 95% and sensitivity below 0.1 nrad/s for a 1-s measurement.

Its upgrade GINGER is aiming at a Lense–Thirring test at the level of 1%,[3–5] *the first measurement* of a GR dynamic effect of the gravitational field on the surface of the Earth (not considering gravitational redshift). Though not in free fall, it would be a direct local measurement, independent of the global distribution of the gravitational field and not an average value, as in the case of space-based experiments. The preparatory phase has provided solutions to scale-factor stabilization[6–8] and data analysis in order to eliminate the nonlinearity induced by laser dynamics.[9] GINGER has to push the relative sensitivity to the Earth's rotation-rate measurement from three parts in 10^9 to one part in 10^{12}. This is equivalent to reaching a sensitivity level of 10^{-16} rad/s.[c]

RLGs have been met with great interest in fundamental physics to study the properties of the gravito-magnetic field and have recently been empoyed for Lorentz-symmetry tests by Jay Tasson and Max Trostel.[10] The RLG array measures the total angular rotation rate, which is the sum of the cinematic Earth rotation rate $\Omega_\oplus$, and the GR terms. Since the international system IERS measures $\Omega_\oplus$, and the de Sitter term is very well known, it is possible to evaluate the Lense–Thirring one. The problem of the Lense–Thirring test is a very general one, similar to any effect induced by geophysical phenomena. The same apparatus is suitable to observe the signal at the sidereal-day frequency, which contains information on Lorenz violation.[10]

Figure 1 shows a pictorial view of GINGER with three independent RLGs and the expected sensitivity as a function of the side length of the square RLGs; based on the present sensitivity, it is assumed that the losses depend only on the four mirrors, the minimum total losses are 6 ppm.[d] From the above, it is clear that RLGs with 10 m-long sides could completely satisfy the GINGER target. The installation inside the Gran Sasso laboratory has been studied, where rings with sides of 5–6 m in length are feasible. At present, we are discussing installation inside the SAR-GRAV underground laboratory, which, being under construction, could accept larger rings.

---

[c]It is important to note that GINGER requires a very high accuracy, and it is advantageous that its apparatus can provide two different measurement techniques: it could act as active or passive, providing an important potential check of the systematics. Moreover, other different laser wavelengths could be used.
[d]At present, the state of the art for mirrors is about 1.5 ppm total losses for the mirrors, 1 ppm scattering, and 0.5 ppm transmission.

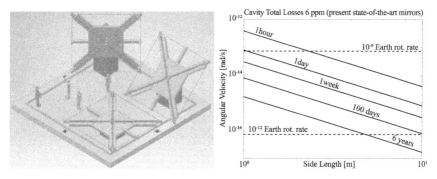

Fig. 1: Pictorial view of the GINGER project, and expected sensitivity as a function of side length of the square RLGs of the array for integration times ranging from one hour to six years.

In summary, GINGER is a project to build an array of RLGs for the terrestrial detection of the Lense–Thirring effect at 1% precision; it is an INFN project, which involves the INFN sections and universities of Pisa, Legnaro, and Naples. Large experimental work has been pursued toward GINGER, focusing on:[6–9] geometrical scale-factor control, optimization of individual RLG orientations, and signal reconstruction following the laser dynamics. GINGERINO, realized for test purposes inside the Gran Sasso undergorund laboratory, has shown the advantage of an underground location,[11] shielded from atmospheric and thermal variations. Presently, its construction is under discussion, and the underground laboratories of Gran Sasso and the future SAR-GRAV are possible locations.

References

1. G. Alexandre, ed., *The Sagnac Effect: 100 Years Later*, Dossier Sommaire, Comptes Rendus Physique, **15**, iii–iv, (2014).
2. K.U. Schreiber and J.-P.R. Wells, Rev. Sci. Instrum. **84**, 041101 (2013).
3. F. Bosi *et al.*, Phys. Rev. D **84**, 122002 (2011).
4. A. Tartaglia *et al.*, Eur. Phys. J. Plus **132**, 73 (2017).
5. A.D.V. Di Virgilio Eur. Phys. J. Plus **132**, 157 (2017).
6. R. Santagata *et al.*, Class. Quant. Grav. **32**, 055013 (2015).
7. J. Belfi *et al.*, Class. Quant. Grav. **31**, 225003 (2014).
8. J. Belfi *et al.*, arXiv:1902.02993.
9. A.D.V. Di Virgilio *et al.*, Eur. Phys. J. C **79**, 573 (2019).
10. J. Tasson and M. Trostel, these proceedings.
11. J. Belfi *et al.*, Appl. Opt. **57**, 5844 (2018).

Constraints on Lorentz-Invariance Violations from Gravitational-Wave Observations

Anuradha Samajdar

Nikhef, Amsterdam, 105 Science Park, 1098 XG, The Netherlands

On behalf of the LIGO Scientific Collaboration and Virgo Collaboration

Using a deformed dispersion relation for gravitational waves, Advanced LIGO and Advanced Virgo have been able to place constraints on violations of local Lorentz invariance as well as the mass of the graviton. We summarize the method to obtain the current bounds from the ten significant binary-black-hole detections made during the first and second observing runs of the above detectors.

1. Introduction

The year 2015 saw the advent of gravitational-wave (GW) astronomy with GW150914,[1] the first directly detected GW signal from a binary-black-hole (BBH) merger. Reference 2 performed tests on strong-field gravity in the highly dynamical regime of General Relativity (GR) finding no statistically significant violations of GR. Since then, ten significant BBH signals have been detected, in addition to a binary-neutron-star (BNS) signal.[3] The first constraints on local Lorentz violation (LV) using real GW data were reported in Ref. 4. These bounds have been revised recently and reported in Ref. 5. These bounds, however, rely on propagation effects and therefore do not directly probe the dynamical regime of gravity.

In these proceedings, we give a brief overview of the method to constrain LV in Sec. 2. We summarize the results of these efforts with some concluding remarks in Sec. 3.

2. Method

GWs propagating in GR are nondispersive and travel with the speed of light. Following Refs. 6,7, we adopt the generic dispersion relation

$$E^2 = p^2 c^2 + A_\alpha p^\alpha c^\alpha. \tag{1}$$

This is a LV dispersion relation for $\alpha > 0$ in which LV is characterized by A_α. In the special case $\alpha = 0$, we may parameterize the additional term in Eq. (1) as $A_0 = m_g^2 c^4$, m_g being the mass of the graviton. Examples of LV theories for specific forms of Eq. (1) include Doubly Special Relativity[8] for $\alpha = 3$ and Hořava–Lifshitz theory[9] for $\alpha = 4$, cf. Refs. 4,5 for more examples and corresponding references. As noted in Ref. 4, a combination of values of α and the sign of A_α can indicate whether the speed of GWs is subluminal or superluminal.

In the presence of dispersion, the low (high)-frequency components of a GW signal travel slower (faster) and result in an overall offset in arrival times at the detector leading to a frequency-dependent shift in the phasing. In frequency domain (FD), the total phase is then given by $\Psi(f) = \Psi_{GR}(f) + \Psi_\alpha(f)$, where $\Psi_{GR}(f)$ is the phasing obtained from GR predictions, and $\Psi_\alpha(f)$ denotes the phase shift following from the dispersion. The waveform model in FD used in our analysis is constructed by $\tilde{h}(f) = \mathcal{A}(f)e^{-i\Psi}$. We associate a length-scale $\lambda_A = hc|A_\alpha|^{1/(\alpha-2)}$ with the LV parameter, where h is Planck's constant and c is the speed of light. The quantity λ_A may be thought of as a screening length corresponding to an effective gravitational potential. In terms of λ_A, the phasing relations are given by

$$\Psi_\alpha(f) = \begin{cases} \text{sign}(A_\alpha)\frac{\pi D_1}{|\lambda_A|}\ln(\pi\mathcal{M}f), & \text{if } \alpha = 1, \\ -\text{sign}(A_\alpha)\frac{\pi}{(1-\alpha)}\frac{D_\alpha}{|\lambda_A|^{2-\alpha}}\frac{f^{\alpha-1}}{(1+Z)^{1-\alpha}}, & \text{if } \alpha \neq 1. \end{cases} \tag{2}$$

In the above equation, $\mathcal{M}$ is the detector-frame chirp mass of the binary system, a combination of component masses given by $\mathcal{M} = (m_1 m_2)^{3/5}/(m_1 + m_2)^{1/5}$, m_1 and m_2 being the component masses, f is the frequency component, Z denotes the redshift to the source, and D_α is a cosmological distance. See Refs. 5,6 for more details.

The analysis carried out in the following section is based on a Bayesian framework which incorporates Bayes' theorem $p(\vec{\theta}|d) = p(d|\vec{\theta})p(\vec{\theta})/p(d)$, where $\vec{\theta}$ refers to a parameter set, d refers to the data, $p(\vec{\theta}|d)$ refers to the posterior probability density obtained on $\vec{\theta}$ from the likelihood calculated from the data $p(d|\vec{\theta})$ and the *a priori* probability density given by $p(\vec{\theta})$; $p(d)$ is a normalizsation constant. The information learned from the data is folded in the likelihood, which takes the following form

$$p(d|\vec{\theta}) \propto \exp\left[-\tfrac{1}{2}(d - h|d - h)\right]. \tag{3}$$

In the presence of a GW signal, the data output from the detector is $d = h(t) + n(t)$, where $h(t)$ is the GW signal and $n(t)$ is the noise. For our analysis, the likelihood integral is computed in FD by including the LV-deformed phase in the model waveform. For a value of α, this enables us to obtain a posterior probability-density function on the parameter A_α leading to a constraint on LV.

3. Results

Being a propagation effect, the strongest constraints come from events located at larger luminosity distances. The bounds obtained from the catalogue of ten sources are presented in Fig. 1. The current bounds obtained from combining all sources lead to an improvement in previously reported bounds[4] by factors of up to 2.4 as reported in Ref. 5.

From combining these sources, the mass of the graviton has been constrained to $m_g \leq 5.0 \times 10^{-23}\,\mathrm{eV/c^2}$ at 90% confidence.

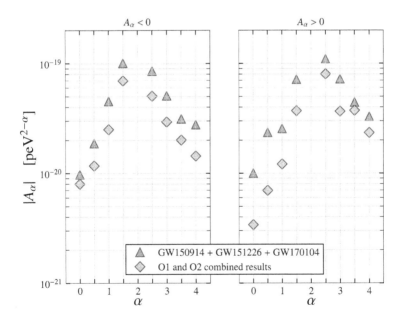

Fig. 1: Upper bounds on A_α (90% credibility) from the BBH detections GW150914, GW151226 and GW170104 (triangles) and those from combining all ten significant BBH detections (diamonds) in O1 and O2. This figure was taken from Ref. 5 and converted to grayscale.

Acknowledgments

The authors gratefully acknowledge the support of the United States National Science Foundation (NSF) for the construction and operation of the LIGO Laboratory and Advanced LIGO as well as the Science and Technology Facilities Council (STFC) of the United Kingdom, the Max-Planck-Society (MPS), and the State of Niedersachsen/Germany for support of the construction of Advanced LIGO and construction and operation of the GEO600 detector. Additional support for Advanced LIGO was provided by the Australian Research Council. The authors gratefully acknowledge the Italian Istituto Nazionale di Fisica Nucleare (INFN), the French Centre National de la Recherche Scientifique (CNRS) and the Foundation for Fundamental Research on Matter supported by the Netherlands Organisation for Scientific Research, for the construction and operation of the Virgo detector and the creation and support of the EGO consortium. The authors also gratefully acknowledge research support from these agencies as well as by the Council of Scientific and Industrial Research of India, the Department of Science and Technology, India, the Science & Engineering Research Board (SERB), India, the Ministry of Human Resource Development, India, the Spanish Agencia Estatal de Investigación, the Vicepresidència i Conselleria d'Innovació, Recerca i Turisme and the Conselleria d'Educació i Universitat del Govern de les Illes Balears, the Conselleria d'Educació, Investigació, Cultura i Esport de la Generalitat Valenciana, the National Science Centre of Poland, the Swiss National Science Foundation (SNSF), the Russian Foundation for Basic Research, the Russian Science Foundation, the European Commission, the European Regional Development Funds (ERDF), the Royal Society, the Scottish Funding Council, the Scottish Universities Physics Alliance, the Hungarian Scientific Research Fund (OTKA), the Lyon Institute of Origins (LIO), the Paris Île-de-France Region, the National Research, Development and Innovation Office Hungary (NKFIH), the National Research Foundation of Korea, Industry Canada and the Province of Ontario through the Ministry of Economic Development and Innovation, the Natural Science and Engineering Research Council Canada, the Canadian Institute for Advanced Research, the Brazilian Ministry of Science, Technology, Innovations, and Communications, the International Center for Theoretical Physics South American Institute for Fundamental Research (ICTP-SAIFR), the Research Grants Council of Hong Kong, the National Natural Science Foundation of China (NSFC), the Leverhulme Trust, the Research Corporation, the Ministry of Science and Technology (MOST), Taiwan and the Kavli Foundation. The authors gratefully acknowledge the support of the NSF, STFC, MPS, INFN, CNRS and the State of Niedersachsen/Germany for provision of computational resources.

References

1. LIGO Scientific and Virgo Collaborations, B.P. Abbott *et al.*, Phys. Rev. Lett. **116**, 061102 (2016).
2. LIGO Scientific and Virgo Collaborations, B.P. Abbott *et al.*, Phys. Rev. Lett. **116**, 221101 (2016); Erratum: Phys. Rev. Lett. **121**, 129902.
3. LIGO Scientific and Virgo Collaborations, B.P. Abbott *et al.*, arXiv:1811.12907.
4. LIGO Scientific and Virgo Collaborations, B.P. Abbott *et al.*, Phys. Rev. Lett. **118**, 221101 (2017); Erratum: Phys. Rev. Lett. **121**, 129901 (2018).
5. LIGO Scientific and Virgo Collaborations, B.P. Abbott *et al.*, arXiv:1903.04467.
6. S. Mirshekari, N. Yunes, and C.M. Will, Phys. Rev. D **85**, 024041 (2012).
7. N. Yunes, K. Yagi, and F. Pretorius, Phys. Rev. D **94**, 084002 (2016).
8. G. Amelino-Camelia, Nature **418**, 34 (2002).
9. P. Hořava, Phys. Rev. D **79**, 084008 (2009).

Mining the Data Tables for Lorentz and CPT Violation

N.E. Russell

Physics Department, Northern Michigan University,Marquette, MI 49855, USA

In this conference proceedings, some comments on the present status and recent growth of efforts to find Lorentz and CPT violation are given by extracting metrics from the annually updated *Data Tables for Lorentz and CPT Violation*. They reveal that tests span all the sectors of particle and gravitational physics, and have shown remarkable and consistent growth. Through numerous innovations and refinements in experiments, a large body of data has been amassed with ever-increasing precisions. The Tables are available through the Cornell preprint archive at arXiv:0801.0287.

The first edition of the *Data Tables for Lorentz and CPT Violation* appeared in the Proceedings of the Fourth Meeting on Lorentz and CPT Symmetry and on the Cornell preprint server arXiv.org, where it has been updated each year.[1] A July 2010 edition appeared in Reviews of Modern Physics.[2] The purpose of the Tables is to maintain a complete list of measurements of coefficients for Lorentz and CPT violation, to summarize the maximal attained sensitivities in high-activity sectors, and to provide information about properties and definitions relevant to the study of these fundamental symmetries.

With the Tables being in existence for more than ten years, and on the occasion of this Eighth Meeting on Lorentz and CPT Symmetry being held more than 20 years since the first such meeting in 1998, it is a good time to take a retrospective look at the extraordinary theoretical and experimental efforts that have been made by hundreds of researchers since the 1990s to find evidence of Lorentz violation in nature.

The burgeoning number of publications placing limits on Lorentz and CPT symmetry can be traced in the annual updates to the Tables. The upper plot in Fig. 1 shows the number of references in each edition of the Tables growing five-fold from 62 in 2008 to 292 in 2019. Another measure is the page count, which has grown about ten-fold from twelve to 115 over the same period.

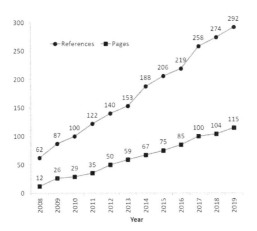

Fig. 1: The number of bibliographic references and the page count for the twelve editions of the *Data Tables for Lorentz and CPT Violation* up to January 2019.

The Standard-Model Extension, or SME, categorizes Lorentz violation in the behavior of all known particles and their interactions,[3] including gravitational ones.[4] It is the result of considering effective field theory with Lorentz and CPT violation, leading to a general framework incorporating known physics but also admitting violations of these symmetries. The Tables list measurements of the coefficients controlling Lorentz- and CPT-breaking operators of all mass dimensions in the framework.

Figure 2 indicates the distribution of efforts across the broad scope of the field. The size of each slice of the pie chart is set by the number of publications placing limits on coefficients for Lorentz and CPT violation. The QED sectors account for more than half of the chart. Limits in the photon sector, the largest one, have been reported in 77 publications to date, with the electron, proton, and neutron sectors represented by 48, 34, and 27 publications, respectively. Rounding out the QED sector, eleven papers have presented limits on Lorentz-breaking couplings between matter and photons.

In the neutrino sector, Fig. 2 shows 40 papers are referenced. In fact, the most recent edition of the Tables has 27 pages of neutrino-sector limits. The quark sector is also well represented, with 23 publications. In comparison, the lepton, electroweak, and gluon sectors have fewer publications so far, reflecting the corresponding experimental challenges and, possibly, revealing some of the many areas where Lorentz and CPT searches could

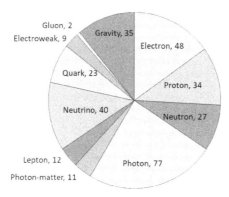

Fig. 2: The number of Data Table references presenting limits in each of the sectors of the Standard-Model Extension.

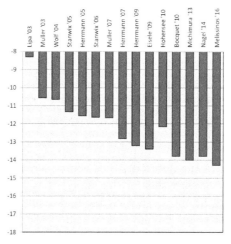

Fig. 3: Evolution of sensitivities to $(\tilde{\kappa}_{o+})^{YZ}$, one of the 19 coefficients governing Lorentz-violating operators of mass dimension four in the photon sector. The scale is logarithmic, with the first limit placed at a few parts in 10^9. The first-named author and the year of publication is listed for each result.[5]

be very profitable. Moving into curved spacetime, the gravitational sector is well represented with 35 publications to date.

Experimentalists have risen to the challenge of finding the elusive effects of Lorentz violation by steadily improving the technology over a period of years. As an example of this, Fig. 3 shows the evolution in sensitivity to the coefficient $(\tilde{\kappa}_{o+})^{YZ}$. The first limit on this dimensionless coefficient was placed in 2003, and up to the present it has been improved 13 times. The sequence of results, originating in labs on three continents, spans more than a decade and shows the experimental reach improving by six orders of magnitude.

At present, the Tables show no evidence of Lorentz or CPT violation. They also show that sectors with unmeasured or very weakly constrained coefficients exist. The potential for finding evidence of Plank-scale physics is of course a central motivator for the field, and phenomenological and experimental efforts continue to grow. As experimental technology improves, the prospect of violations being revealed remains tantalizing.

References

1. *Data Tables for Lorentz and CPT Violation,* V.A. Kostelecký and N. Russell, arXiv:0801.0287v1 (2008 edition) to arXiv:0801.0287v12 (2019 edition); *id.,* in V.A. Kostelecký, ed., *CPT and Lorentz Symmetry IV*, World Scientific, Singapore, 2008.
2. V.A. Kostelecký and N. Russell, Rev. Mod. Phys. **83**, 11 (2011).
3. D. Colladay and V.A. Kostelecký, Phys. Rev. D **55**, 6760 (1997); Phys. Rev. D **58**, 116002 (1998).
4. V.A. Kostelecký, Phys. Rev. D **69**, 105009 (2004).
5. J.A. Lipa *et al.*, Phys. Rev. Lett. **90**, 060403 (2003); H. Müller *et al.*, Phys. Rev. Lett. **91** 020401 (2003); P. Wolf *et al.*, Phys. Rev. D **70**, 051902 (2004); P.L. Stanwix *et al.*, Phys. Rev. Lett. **95**, 040404 (2005); S. Herrmann *et al.*, Phys. Rev. Lett. **95**, 150401 (2005); P.L. Stanwix *et al.*, Phys. Rev. D **74**, 081101(R) (2006); H. Müller *et al.*, Phys. Rev. Lett. **100**, 031101 (2008); S. Herrmann *et al.*, in V.A. Kostelecký, ed., *CPT and Lorentz Symmetry IV*, World Scientific, Singapore, 2008; S. Herrmann *et al.*, Phys. Rev. D **80**, 105011 (2009); Ch. Eisele, A.Yu. Nevsky, and S. Schiller, Phys. Rev. Lett. **103**, 090401 (2009); M.A. Hohensee *et al.*, Phys. Rev. D **82**, 076001 (2010); J.-P. Bocquet *et al.*, Phys. Rev. Lett. **104**, 241601 (2010); Y. Michimura *et al.*, Phys. Rev. Lett. **110**, 200401 (2013); M. Nagel *et al.*, Nature Commun. **6**, 8174 (2015); V.A. Kostelecký, A.C. Melissinos, and M. Mewes, Phys. Lett. B **761**, 1 (2016).

Search for Lorentz Violation Using High-Energy Atmospheric Neutrinos in IceCube

Carlos A. Argüelles

Massachusetts Institute of Technology, Cambridge, MA 02139, USA

On behalf of the IceCube Collaboration

High-energy atmospheric neutrinos observed by the IceCube Neutrino Observatory are extremely sensitive probes of Lorentz violation. Here, we report the result of analyzing two years of IceCube data in the search for Lorentz violation. This analysis places some of the strongest constraints on Lorentz violation when considering higher-dimensional operators.

1. Lorentz violation effects on high-energy neutrinos

Neutrino flavor information is one of the most sensitive observables for Lorentz violation (LV). This is due to the fact that neutrino flavor changes are produced by the interference between different neutrino states. It is only through the study of neutrino flavor morphing that we have been able to infer the existence of neutrino masses; direct neutrino mass measurements have so far yielded null results.[1] The power of neutrino interferometry has been demonstrated in the study of neutrino and antineutrino oscillation parameters, which provides a strong test for CPT symmetry.[2,3] The neutrino to antineutrino mass-squared differences are limited to $\left|\Delta m_{31}^2 - \Delta \overline{m}_{31}^2\right| < 0.8 \times 10^{-3}\,\text{eV}^2$ while the neutral-kaon mass difference[2] is only bounded to be $\left|m^2\left(K^0\right) - m^2\left(\bar{K}^0\right)\right| < 0.25\,\text{eV}^2$.

We can incorporate the effect of LV into neutrino oscillations by considering the following hamiltonian:[4]

$$H \sim \frac{m^2}{2E} + \mathring{a}^{(3)} - E \cdot \mathring{c}^{(4)} + E^2 \cdot \mathring{a}^{(5)} - E^3 \cdot \mathring{c}^{(6)} \cdots, \qquad (1)$$

where the first term is responsible for the measured neutrino oscillations and the latter terms encode the effects of LV. The a terms are CPT-odd, the c terms CPT-even, and each index denotes the dimension of the operator that produced this term. Each of these coefficients is a matrix with flavor indices. In this work, we restrict ourselves to study the μ–τ sector; see

Fig. 1 (left). At high energies, where $\frac{m^2}{2E}$ is negligible, the ν_τ appearance probability is given by[5]

$$P\left(\nu_\mu \to \nu_\tau\right) \sim \frac{\left| \overset{\circ}{a}_{\mu\tau}^{(d)} - \overset{\circ}{c}_{\mu\tau}^{(d)} \right|^2}{\rho_d^2} \sin^2\left(\rho_d L \cdot E^{d-3}\right), \qquad (2)$$

where we have introduced the LV strength

$$\rho_d \equiv \sqrt{\left(\overset{\circ}{a}_{\mu\mu}^{(d)}\right)^2 + \operatorname{Re}\left(\overset{\circ}{a}_{\mu\tau}^{(d)}\right)^2 + \operatorname{Im}\left(\overset{\circ}{a}_{\mu\tau}^{(d)}\right)^2}; \qquad (3)$$

a similar relation holds for the c terms. As can be seen in Eq. (2), the appearance probability increases with neutrino energy and traveled distance. In this work, we focus on the effects of LV on neutrino flavors, but it is worthwhile to note that other LV effects can be studied with atmospheric neutrinos. LV can also affect the production of neutrinos in cosmic-ray air showers,[6,7] but the use of these effects to probe LV is expected to be less sensitive than neutrino interferometry.

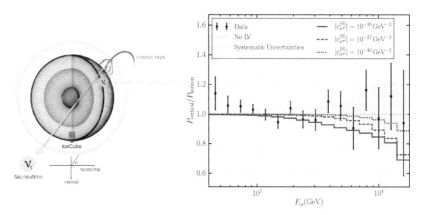

Fig. 1: Illustration of the effects of LV in high-energy atmospheric neutrinos. Left: Cartoon of the analysis. A muon neutrino produced in a cosmic-ray air shower travels through the Earth and converts to a tau neutrino due to interactions with the all-permeating LV field (light gray background arrows). Right: The double-ratio of vertical to horizontal events as a function of reconstructed muon energy. A horizontal line at one corresponds to the Standard Model, other lines show the expected double ratio for various values of the LV dimension six coefficient. Data points with statistical error bars are shown as well. The gray band is the total uncorrelated systematic uncertainty.

2. IceCube and high-energy atmospheric neutrinos

The IceCube Neutrino Observatory is a gigaton-scale neutrino detector located in Antarctica.[8] IceCube currently has the largest sample of neutrinos above a TeV due to its large effective area. These neutrinos are produced by cosmic rays interacting in the Earth's atmosphere. At TeV energies, atmospheric neutrinos are primarily produced by kaon decay. At higher energies, neutrinos from charmed hadrons are expected to be important, but this component has yet to be observed. Finally, for energies above $\sim 100\,\text{TeV}$, extraterrestrial neutrinos dominate the flux and are expected to be the most sensitive to LV.[9] Since we are looking for a flavor-changing effect, we restrict the analysis to reconstructed muon energies below 18 TeV. In this regime, atmospheric neutrinos, whose flavor composition is well predicted, dominate the flux. The effect of the LV dimension six operator is illustrated in Fig. 1 (right). This shows the double ratio of horizontal-to-vertical data-to-prediction as a function of muon energy. As the LV flavor-violating coefficient is increased, muon neutrinos are starting to disappear at the highest energies.

3. Results

We have performed a search for muon-neutrino disappearance induced by LV using two years of IceCube data described in Ref. 10. This data set corresponds to through-going muons and was previously used to search for evidence of a high-energy astrophysical neutrino component. This set contains 34975 events with a 0.1% atmospheric-muon contamination spanning an energy range from 400 GeV to 20 TeV in muon energy. Using this data, we find no evidence for anomalous disappearance and place bounds on the flavor-violating effects defined in the LV Standard-Model Extension. The obtained constraints are among the strongest bounds on LV and are particularly important for higher-dimensional operators.[11] In Fig. 2, the results for the dimension-six operator are shown. The left panel plots the ratio of diagonal to total LV strength as a function of the LV strength. When the diagonal component dominates, no flavor change is introduced, and thus no constraint is obtained. The constraint is strongest when the diagonal component vanishes, which is the case for maximal flavor violation. This limits the lower and upper parts of the constrained region. The smallest strengths are limited by the statistical uncertainty of high-energy atmospheric neutrinos, and the largest strengths are limited by the uncertainty of the absolute normalization of the atmospheric flux. To compare with

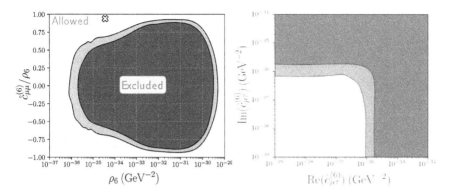

Fig. 2: Constraints on LV dimension six operators. Left: Constraints in terms of the LV strength (horizontal axis) and the ratio of diagonal to total strength (vertical axis). The best-fit point of the analysis is marked by the standard gray cross. Right: Constraints assuming a maximally violating flavor texture, i.e., diagonal elements set to zero, as a function of the real and imaginary parts of the off-diagonal elements. In both panels, points inside the gray (black) patch are excluded at 90% (99%) c.l.

other results in the literature,[12] we report our results for the maximum flavor-violation scenario in the right panel of Fig. 2. Finally, even though these results have been shown in terms of interactions between neutrinos and a hypothetical LV field, they can be recast to other scenarios, such as neutrino–dark-matter interactions.[13–15]

References

1. Particle Data Group, M. Tanabashi *et al.*, Phys. Rev. D **98**, 030001 (2018).
2. G. Barenboim, C.A. Ternes, and M. Tórtola, Phys. Lett. B **780**, 631 (2018).
3. G. Barenboim and J. Salvado, Eur. Phys. J. C **77**, 766 (2017).
4. V.A. Kostelecký and M. Mewes, Phys. Rev. D **85**, 096005 (2012).
5. M.C. Gonzalez-Garcia *et al.*, Phys. Rev. D **71**, 093010 (2005).
6. J.S. Díaz *et al.*, Phys. Rev. D **94**, 085025 (2016).
7. B. Altschul, Phys. Rev. D **88**, 076015 (2013).
8. IceCube Collaboration, M.G. Aartsen *et al.*, J. Instrum. **12**, P03012 (2017).
9. C. Argüelles, T. Katori, and J. Salvado, Phys. Rev. Lett. **115**, 161303 (2015).
10. IceCube Collaboration, M.G. Aartsen *et al.*, Phys. Rev. Lett. **115**, 081102 (2015).
11. IceCube Collaboration, M.G. Aartsen *et al.*, Nature Phys. **14**, 961 (2018).
12. Super-Kamiokande Collaboration, K. Abe *et al.*, Phys. Rev. D **91**, 052003 (2015).
13. N. Klop and S. Ando, Phys. Rev. D **97**, 063006 (2018).
14. Y. Farzan and S. Palomares-Ruiz, Phys. Rev. D **99**, 051702 (2019).
15. F. Capozzi, I.M. Shoemaker, and L. Vecchi, J. Cosmol. Astropart. Phys. **1807**, 004 (2018).

Lorentz Violation and Radiative Corrections in Gauge Theories

A.F. Ferrari

Centro de Ciências Naturais e Humanas, Federal do ABC,
Avenida dos Estados, 5001, Santo André - SP, 09210-580, Brazil

Various studies have already considered radiative corrections in Lorentz-violating models unveiling many instances where a minimal or nonminimal operator generates, via loop corrections, a contribution to the photon sector of the Standard-Model Extension. However, an important fraction of this literature does not follow the widely accepted conventions and notations of the Standard-Model Extension, and this obscures the comparison between different calculations as well as possible phenomenological consequences. After reviewing some of these works, we uncover one example where a well defined loop correction to the k_F coefficient, already presented in the literature, allows us to improve the bounds on one specific coefficient of the fermion sector of the Lorentz-violating QED extension.

The Standard-Model Extension (SME)[1,2] is understood as an effective field theory based on the internal symmetries and field content of the Standard Model, but incorporating Lorentz violation (LV) in a very general way. A broad experimental program has used the SME framework to obtain stringent bounds on possible LV operators using different experiments and astrophysical observations.[3] The most studied and well constrained sector of the SME is the LV extension to the Maxwell theory, defined in its most general form in Ref. 4. In particular, the minimal photon sector involves only two LV coefficients, k_{AF} and k_F. Both terms in general induce birefringence in the vacuum leading to very strong constraints from astrophysics: of order 10^{-43} GeV for k_{AF} and 10^{-37} for the birefringent components of k_F.[3]

If taken as an effective parametrization for LV effects to be searched for in the laboratory, the SME might be understood as a strictly tree-level theory. From the theoretical viewpoint, however, it motivates many interesting studies concerning the consistency, as a full quantum field theory (QFT), of theories in which one of the central aspects of the QFT formalism—Lorentz symmetry—is in some sense violated. Different field-theoretical aspects, such as renormalization, the structure of asymptotic states, and others, have already been investigated.[5–9]

Radiative corrections, in particular, may lead to results of direct phenomenological interest: some set of LV coefficients may generate or contribute to other sets of LV coefficients via loop corrections. An early example was discussed in detail in Ref. 10: the b term

$$b_\mu \bar{\psi} \gamma^\mu \gamma_5 \psi, \tag{1}$$

which is a LV correction to the fermion propagator, will induce a one-loop correction to the quadratic photon lagrangian of the form

$$C e^2 \epsilon^{\mu\nu\lambda\rho} b_\mu A_\nu \partial_\lambda A_\rho, \tag{2}$$

e being the fermion charge and C a constant. In the SME notation, this amounts to the generation of a minimal k_{AF} term with $k_{AF} \sim b$ or to a correction to a k_{AF} term already present at tree level. The phenomenological interest in such a result is that k_{AF} can be strongly constrained by photon vacuum birefringence, and these constraints could be translated to b^μ. However, it was readily recognized that the result presented in Eq. (2) is anomalous: the loop integral turns out to be finite but ambiguous, its result being dependent on the regularization scheme used to calculate it (for a recent discussion of this problem, see Ref. 11). In this case, it is certainly not possible to use loop corrections to obtain sound phenomenological conclusions.

Many different instances of radiative generation of LV operators have been presented in the literature, and we might wonder whether finite and well-defined corrections can be calculated in some cases, and whether stringent bounds obtained in one sector of the SME might be transferred to other, perhaps not so well bounded sectors. This requires some work since many of the reported calculations do not follow the now standard SME notations.

As an example, higher-derivative corrections to the photon sector originating from (1) where presented in Ref. 12 as

$$\mathcal{L}_{eff} \supset \frac{e^2}{24\pi^2 m^2} \epsilon^{\beta\mu\nu\rho} b_\beta A_\mu \Box F_{\nu\rho}, \tag{3}$$

corresponding in SME notation to the generation of the dimension five coefficient

$$(\hat{k}_{AF}^{(5)})_\kappa{}^{\alpha\beta} = \frac{e^2}{48\pi^2 m^2} b_\kappa \eta^{\alpha\beta}. \tag{4}$$

This is a finite result, free of ambiguities. Unfortunately, this form of $\hat{k}_{AF}^{(5)}$ does not modify free propagation of photons at leading order, so no interesting bounds can be obtained from this result at the moment.

Looking at one-loop corrections involving higher orders in b, we may also obtain finite and well-defined results, such as[13,14]

$$\mathcal{L}_{eff} \supset -\frac{e^2}{6\pi^2 m^2}\, b^\mu b_\lambda\, F_{\mu\nu} F^{\lambda\nu}, \tag{5}$$

which, written in the SME notation, corresponds to

$$(k_F^{(4)})^{\mu\nu\alpha\beta} = -\frac{e^2}{6\pi^2 m^2}\left(b^\mu b^\alpha \eta^{\nu\beta} - b^\nu b^\alpha \eta^{\mu\beta} - b^\mu b^\beta \eta^{\nu\alpha} + b^\nu b^\beta \eta^{\mu\alpha}\right). \tag{6}$$

At first sight, one might not expect to find competitive constraints from this expression, since it is of second order in b. However, birefringent components of k_F are constrained at the order of 10^{-37}, and so this result would provide a bound of the order $b^2 < 6\pi^2 m^2/e^2 \times 10^{-37}$ for the b coefficient, corresponding to $|b_p^\mu| < 3\times10^{-15}$ GeV for protons and $|b_e^\mu| < 1.5\times10^{-20}$ GeV for electrons, for example. These are not better (but also not much worse) than the constraints already found *for the space components* of b^μ, as measured in the Sun-centered reference frame.[3] On the other hand, the temporal component b^T is not so well constrained: the best bounds are of order $|(b^T)_p| < 7\times10^{-8}$ GeV for protons and $|(b^T)_e| < 10^{-15}$ GeV for electrons. Therefore, the radiative correction presented in Eq. (6) can translate the stringent constraints on birefringent components of k_F to competitive bounds on the temporal components b^T. It remains to check that this particular form of $k_F^{(4)}$ does indeed induce birefringence. The easiest way to do this is by using the parametrization of birefringent components of k_F in terms of the ten k^a coefficients as given in Ref. 15: one may easily verify that Eq. (6) corresponds to non-vanishing coefficients $k^a = e^2 b^2/6\pi^2 m^2$ for $a = 3, 4$.

The end result is therefore the constraint

$$|b_T| < \pi m/e\sqrt{6\times10^{-37}}, \tag{7}$$

for the temporal component, in the Sun-centered frame, of the b coefficient for a given fermion, depending on its mass m. For example, we have

$$|(b_T)_p| < 3\times10^{-15}\,\text{GeV} \tag{8}$$

for protons and

$$|(b_T)_e| < 1.5\times10^{-20}\,\text{GeV} \tag{9}$$

for electrons.

This result was presented in Ref. 16, together with an extensive study of other instances of radiative corrections in different sectors of the SME. It is an interesting example, where a weakly bounded coefficient for LV can be subjected to a stronger constraint, borrowed from the very well studied photon sector of the SME. We believe that, besides interesting questions regarding theoretical consistency and technical challenges of calculating ambiguous Feynman integrals, the study of radiative corrections might help to fill some of the gaps in the extensive set of searches for LV that have been developed in the last decades using the SME as the fundamental framework.

Acknowledgments

This work was partially supported by Fundação de Amparo à Pesquisa do Estado de São Paulo (FAPESP) and Conselho Nacional de Desenvolvimento Científico e Tecnológico (CNPq) via the following grants: CNPq 304134/2017-1 and FAPESP 2017/13767-9.

References

1. D. Colladay and V.A. Kostelecký, Phys. Rev. D **55**, 6760 (1997).
2. D. Colladay and V.A. Kostelecký, Phys. Rev. D **58**, 116002 (1998).
3. *Data Tables for Lorentz and CPT Violation*, V.A. Kostelecký and N. Russell, 2019 edition, arXiv:0801.0287v12.
4. V.A. Kostelecký and M. Mewes, Phys. Rev. D **80**, 015020 (2009).
5. V.A. Kostelecký, C.D. Lane, and A.G.M. Pickering, Phys. Rev. D **65**, 056006 (2002).
6. D. Colladay and P. McDonald, Phys. Rev. D **79**, 125019 (2009).
7. A. Ferrero and B. Altschul, Phys. Rev. D **84**, 065030 (2011).
8. R. Potting, Phys. Rev. D **85**, 045033 (2012).
9. M. Cambiaso, R. Lehnert, and R. Potting, Phys. Rev. D **90**, 065003 (2014).
10. R. Jackiw and V.A. Kostelecký, Phys. Rev. Lett. **82**, 3572 (1999).
11. B. Altschul, *There is No Ambiguity in the Radiatively Induced Gravitational Chern–Simons Term*, arXiv:1903.10100.
12. J. Leite and T. Mariz, Europhys. Lett. **99**, 21003 (2012).
13. G. Bonneau, L.C. Costa, and J.L. Tomazelli, Int. J. Theor. Phys. **47**, 1764 (2008).
14. T. Mariz, J.R. Nascimento, A.Yu. Petrov, and A.P. Baeta Scarpelli, Eur. Phys. J. C **73**, 2526 (2013).
15. V.A. Kostelecký and M. Mewes, Phys. Rev. Lett. **87**, 251304 (2001).
16. A.F. Ferrari, J.R. Nascimento, and A.Y. Petrov, *Radiative corrections and Lorentz violation*, arXiv:1812.01702.

Status of the KATRIN Neutrino Mass Experiment

Y.-R. Yen

Department of Physics, Carnegie Mellon University, Pittsburgh, PA 15213, USA

On behalf of the KATRIN Collaboration

The KArlsruhe TRItium Neutrino experiment, or KATRIN, is designed to measure the tritium β-decay spectrum with enough precision to be sensitive to the neutrino mass down to 0.2 eV at 90% confidence level. After an initial first tritium run in the summer of 2018, KATRIN is taking tritium data in 2019 that should lead to a first neutrino mass result. The β spectral shape of the tritium decay is also sensitive to four *countershaded* Lorentz-violating, oscillation-free operators within the Standard-Model Extension that may be quite large. The status and outlook of KATRIN to produce physics results, including Lorentz-violation measurements, are discussed.

1. Introduction

The massive nature of neutrinos remains one of the major open questions in physics that cannot be explained by the minimal Standard Model. Among the various ways for directly probing the absolute neutrino mass, β decay is the least model dependent unlike cosmology (cosmological models)[1] or neutrinoless double beta decay (nuclear models and quenching).[2] Kinematics alone defines the β-decay spectrum, from which the electron-antineutrino mass, in the form of the effective mass $m_{\text{eff}}^2(\nu_e) = \Sigma_i |U_{ei}^2| m_i^2$, will determine the spectral-shape modification very near the high-energy endpoint.

A good candidate for very precise β spectral measurements is tritium (T). Tritium (^{3}H) decays into ^{3}He with the emissions of an electron and an antineutrino. The relatively low Q value of 18.3 keV means that the fractional change to the spectrum by the neutrino mass would be large. A high-luminosity tritium source, necessary for high statistics since only a small fraction of the decays are near the endpoint energy, can be designed due to the tritium half-life being 12.3 years.

The previous generation of tritium beta-decay experiments, located at Troitsk and at Mainz, set a limit on the antineutrino mass at $\sim$2 eV at 90% c.l.[3,4] KATRIN is a next-generation tritium experiment that has recently started taking data. The ultimate sensitivity of KATRIN is expected to be ~0.2 eV at 90% c.l., an improvement by an order of magnitude.

2. Experiment description

The KATRIN experiment can achieve its desired sensitivity due to a high-activity, gaseous molecular-tritum (T_2) source and a high-resolution spectrometer. A β electron will be adiabatically guided by magnetic fields to travel down the entire 70-meter beamline to the detector. After the decay takes place in the windowless, gaseous T_2 source, a two-stage transport section eliminates the non-β-electron components (ions and neutral atoms) ahead of the main spectrometer. The main spectrometer, with the Magnetic Adiabatic Collimation combined with an Electrostatic filter design, will select with fine energy resolution (0.93 eV at the endpoint energy of 18.6 keV) only β electrons above a specific energy to pass. This filter adiabatically transforms the β electrons' momenta to the longitudinal direction, the same direction as the precisely set spectrometer voltage, allowing the endpoint silicon PIN diode detector to count essentially just the number of electrons with decay energies above an energy threshold. A more detailed description of the KATRIN experimental setup can be found in Ref. 5. In normal KATRIN operation, an integrated spectrum of β decay is measured.

The measurement-time distribution, i.e., where and how long to scan the β spectrum via the main spectrometer potential setting, is optimized for the best neutrino mass sensitivity. Reference 6 lists the theoretical modification to the β spectrum that KATRIN will need to consider. From the integrated spectrum with those effects considered, the neutrino mass will be one of four fundamental parameters to be fitted, alongside the background, the amplitude, and the endpoint energy, which is also important for Lorentz-violation (LV) searches.

3. Prospects for LV searches with KATRIN

The Standard-Model Extension (SME) provides a framework for small, testable Lorentz and CPT breakdown.[7] In the neutrino sector of the SME, the majority of LV coefficients are expected to be heavily suppressed by the tiny ratio of the electroweak and Planck scales (m_w/m_p) of about 10^{-17}. However, four oscillation-free operators may be examples of the rare *countershaded* LV; these effects may be large but difficult to detect, akin to a shark's different color on its belly and its back. These four operators are renormalizable, flavor blind, and of mass dimension three; other than breaking Lorentz and CPT symmetry, all other physics is conventional.[8]

The corresponding countershaded coefficients to these operators are one timelike $(a_{\text{of}}^{(3)})_{00}$ and three spacelike $(a_{\text{of}}^{(3)})_{1m}$ with $m = 0, \pm 1$. While the 00 term is independent of apparatus location, the spacelike terms depend on χ, ξ, and θ_0 representing the coordinate change to the standard choice of SME frame; for terrestrial experiments like KATRIN, the leading contribution to this transformation arises from the rotation of the Earth around its axis.

The paper by Díaz et al., on *Lorentz violation and beta decay*[8] both analyzed the published Troitsk and Mainz results[3,4] to get the current best limits on two of the oscillation-free operators and provided the road map to figure out the other two operators, which requires binning the data based on sidereal time. The rest of this section will mostly summarize that paper.

For the tritium-decay energy range ΔT near the endpoint energy T_0, i.e., $\Delta T = T_0 - T$ is small, the decay rate can be written as

$$\frac{d\Gamma}{dT} \simeq B + C\left[(\Delta T + k(T_\oplus))^2 - \tfrac{1}{2}m_\nu^2\right], \tag{1}$$

where B is the experimental background rate and C is approximately constant. The SME coefficients contribute to the function $k(T_\oplus)$, which depends on the sidereal time $T_\oplus$:

$$\begin{aligned}
k(T_\oplus) = {} & \tfrac{1}{\sqrt{4\pi}}(a_{\text{of}}^{(3)})_{00} - \sqrt{\tfrac{3}{4\pi}}\cos^2\tfrac{1}{2}\theta_0 \sin\chi \cos\xi\,(a_{\text{of}}^{(3)})_{10} \\
& - \sqrt{\tfrac{3}{2\pi}}\cos^2\tfrac{1}{2}\theta_0\left[\sin\xi\,\text{Im}\big((a_{\text{of}}^{(3)})_{11}e^{i\omega_\oplus T_\oplus}\big)\right. \\
& \left. + \cos\chi \cos\xi\,\text{Re}\big((a_{\text{of}}^{(3)})_{11}^* e^{-i\omega_\oplus T_\oplus}\big)\right],
\end{aligned} \tag{2}$$

where $\text{Re}(a_{\text{of}}^{(3)})_{11} = (a_{\text{of}}^{(3)})_{11}$ and $\text{Im}(a_{\text{of}}^{(3)})_{11} = (a_{\text{of}}^{(3)})_{1-1} \equiv -(a_{\text{of}}^{(3)})_{11}^*$.

Equation 2 shows that $k(T_\oplus)$ will only shift the endpoint energy of the decay spectrum without changing the shape and is independent of the neutrino mass. Without knowing the sidereal time, the harmonic oscillation effects of the $1-1$ and 11 terms on the spectrum will average out. Limits on $(a_{\text{of}}^{(3)})_{00}$ and $(a_{\text{of}}^{(3)})_{10}$ can then be obtained from the limit on the potential endpoint energy shift.

By conservatively taking $< 5\,\text{eV}$ to be the Troitsk and Mainz sensitivity, $|(a_{\text{of}}^{(3)})_{00}| < 2 \times 10^{-8}\,\text{GeV}$ and $|(a_{\text{of}}^{(3)})_{10}| < 5 \times 10^{-8}\,\text{GeV}$ limits were set. For comparison, the neutrinoless double beta-decay experiment EXO-200 also searched for $(a_{\text{of}}^{(3)})_{00}$ from the two-neutrino double beta-decay spectrum shape;[9] their limit of $-9.39 \times 10^{-5}\,\text{GeV} < (a_{\text{of}}^{(3)})_{00} < 2.69 \times 10^{-5}\,\text{GeV}$ is significantly worse than what a single β-decay experiment can do.

Reference 8 predicts that 30 days of nominal KATRIN run can improve the limit on $(a_{\text{of}}^{(3)})_{00}$ by two orders of magnitude. A KATRIN analysis that considers sidereal time can set the first limits on $(a_{\text{of}}^{(3)})_{11}$ and $(a_{\text{of}}^{(3)})_{1-1}$.

4. Outlook

An initial tritium injection June 5th–18th, 2018 resulted in a successful "First Tritium" run of 81 hours worth of tritium data. The data taken from March until May of 2019 with a higher concentration of tritium, named "KNM1," will be the first KATRIN run to be used for a neutrino mass result. With some off-season adjustments, KATRIN plans to operate at the nominal settings starting in 2020. A three-year run spanning five calendar years should allow KATRIN to reach the design sensitivity of 0.2 eV.

The timeline for a KATRIN LV result will depend on the prior release of neutrino mass results. Even if the blinding scheme for the neutrino mass analysis does not mask the endpoint energy result that is needed for the LV analysis, a comprehensive understanding of the endpoint energy systematics will not be available until after the completion of the main neutrino mass analysis. With some strong theoretical motivations to search for LV with the β-spectrum method, KATRIN will likely produce LV results following each neutrino mass result release in the coming years.

Acknowledgments

This work is supported by the DOE Office of Science under award No. #DE-SC0019304.

References

1. Planck Collaboration, N. Aghanim *et al.*, arXiv:1807.06209.
2. M.J. Dolinski, A.W.P. Poon, and W. Rodejohann, arXiv: 1902.04097.
3. C. Kraus *et al.*, Eur. Phys. J. C **40**, 447 (2015).
4. V.N. Assev *et al.*, Phys. Rev. D **84**, 112003 (2011).
5. KATRIN Collaboration, M. Arenz *et al.*, JINST **13**, P04020 (2018).
6. M. Kleesiek *et al.*, Eur. Phys. J. C **79**, 204 (2019).
7. D. Colladay and V.A. Kostelecký, Phys. Rev. D **55**, 6760 (1997); D. Colladay and V.A. Kostelecký, Phys. Rev. D **58**, 116002 (1998); V.A. Kostelecký, Phys. Rev. D **69**, 105009 (2004).
8. J.S. Díaz , V.A. Kostelecký , and R. Lehnert, Phys. Rev. D **88**, 071902 (2013).
9. EXO-200 Collaboration, J.B. Albert *et al.*, Phys. Rev. D **93**, 072001 (2016).

Lorentz-Violating Running of Coupling Constants

A.R. Vieira[*] and N. Sherrill[†]

[*]*Universidade Federal do Triângulo Mineiro, Iturama, MG 38280-000, Brazil*

[†]*Physics Department, Indiana University, Bloomington, IN 47405, USA*

We compute the full vacuum polarization tensor in the minimal QED extension. We find that its low-energy limit is dominated by the radiatively induced Chern–Simons-like term and the high-energy limit is dominated by the c-type coefficients. We investigate the implications of the high-energy limit for the QED and QCD running couplings. In particular, the QCD running offers the possibility to study Lorentz-violating effects on the parton distribution functions and observables such as the hadronic R ratio.

Spacetime anisotropy affects not only clocks and rulers but also masses and couplings. Masses and couplings appearing in the tree-level Lagrangian are just parameters, which acquire corrections due to interactions. These parameters with their quantum corrections are referred to as the physical masses and couplings. Since quantum corrections are modified in the presence of Lorentz-violating effects, it is possible to place limits on Lorentz violation by studying the running of these quantities.

In addition, the coefficient space of the QCD sector of the Standard Model Extension (SME) is comparatively unexplored.[1] Therefore, it is of interest to study how Lorentz violation affects perturbative QCD processes like $e^+e^- \rightarrow$ hadrons, deep inelastic scattering,[2] the Drell–Yan process, and related quantities like parton distribution functions (PDFs).[5] In the following discussion, let us consider the modified lagrangian of a single-flavor fermion:

$$\mathcal{L} = \tfrac{1}{2} i \bar{\psi} \Gamma^\nu \overleftrightarrow{D}_\nu \psi - \bar{\psi} M \psi - \tfrac{1}{4} F_{\mu\nu} F^{\mu\nu} - \tfrac{1}{4} (k_F)_{\kappa\lambda\mu\nu} F^{\kappa\lambda} F^{\mu\nu} + \tfrac{1}{2} (k_{AF})^\kappa \epsilon_{\kappa\lambda\mu\nu} A^\lambda F^{\mu\nu}, \qquad (1)$$

where $D_\mu \equiv \partial_\mu + ieA_\mu$ is the usual covariant derivative, which couples the gauge field with matter,

$$\Gamma^\nu = \gamma^\nu + c^{\mu\nu}\gamma_\mu + d^{\mu\nu}\gamma_5\gamma_\mu + e^\nu + if^\nu\gamma_5 + \tfrac{1}{2}g^{\lambda\mu\nu}\sigma_{\lambda\mu}, \qquad (2)$$

and

$$M = m + m_5\gamma_5 + a_\mu\gamma^\mu + b_\mu\gamma_5\gamma^\mu + \tfrac{1}{2}H_{\mu\nu}\sigma^{\mu\nu}. \tag{3}$$

Considering the full fermion propagator is a difficult task. Instead, we treat the coefficients for Lorentz violation perturbatively by keeping only first-order corrections, which is appropriate given existing experimental results. There is a total of six one-loop diagrams that correct the gauge-boson propagator. Also, the regularization scheme to be applied is a subtle task. As we can see in Eq. (1), some terms contain dimension-specific objects, namely the Levi–Civita symbol and the γ_5 matrix. The inadequate choice of a regulator may cause spurious terms, especially if we are dealing with the finite part of the amplitudes. Therefore, we apply a scheme called implicit regularization.[3] This scheme allows us to stay in four dimensions and does not involve spurious symmetry-breaking terms in the process of renormalization.

After computing the diagrams, we find that the low-energy limit ($p^2 \ll m^2$) of the renormalized vacuum polarization tensor is dominated by the induced Chern–Simons-like term

$$\Pi_{\text{LV}}^{\mu\nu}(p) \approx \tfrac{e^2}{2\pi^2}\epsilon^{\alpha\beta\mu\nu}b_\alpha p_\beta. \tag{4}$$

In computing Eq. (4), we also find contributions from the the c-type and g-type coefficients from Eq. (2). However, they are suppressed relative to the dominant term by powers of p/m. Also, Eq. (4) can be affected by arbitrary and regularization-dependent surface terms. They are null if we require gauge invariance or momentum-routing invariance of the diagrams.[3]

By contrast, in the high-energy limit ($p^2 \gg m^2$) we find

$$\Pi_{\text{LV}}^{\mu\nu}(p) \approx \tfrac{ie^2}{12\pi^2}(p^2 c^{\mu\nu} - p^\mu(c^{\nu p} + c^{p\nu}) + (\mu \leftrightarrow \nu))\left(\ln\tfrac{-p^2}{m^2} - \tfrac{5}{3}\right) + $$
$$+ \tfrac{ie^2}{6\pi^2}\left(\ln\tfrac{-p^2}{m^2} - \tfrac{13}{6}\right)c^{pp}\eta^{\mu\nu} + \tfrac{ie^2}{4\pi^2}c^{pp}\left(\eta^{\mu\nu} - \tfrac{2}{3}\tfrac{p^\mu p^\nu}{p^2}\right), \tag{5}$$

where $c^{p\nu} \equiv c^{\mu\nu}p_\mu$. Equation (5) shows that at high energies the running couplings will effectively depend only on the c-type coefficients. It is also straightforward to show that the Ward identity is fulfilled, $p_\mu\Pi_{\text{LV}}^{\mu\nu}(p) = 0$. If we insert this finite correction into a process, like electron–muon scattering, we can clearly see the c-type coefficients affect the running of the QED coupling. The amplitude of this process takes the form

$$\mathcal{M} = -\tfrac{1}{p^2}[\bar{u}(p_3)\Gamma_\mu(p^2)u(p_1)]e_R^2(p^2)[\bar{u}(p_4)\Gamma^\mu(p^2)u(p_2)], \tag{6}$$

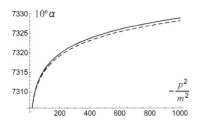
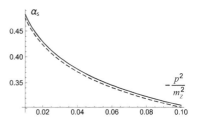

Fig. 1: Running of the QED coupling. The dashed line is the running shifted by Lorentz violation for $\frac{c^{pp}}{p^2} \sim 10^{-2}$.

Fig. 2: Running of the QCD coupling. The dashed line is the running shifted by Lorentz violation for $\frac{c^{pp}}{p^2} \sim 10^{-1}$.

where $p^2 = (p_1 - p_3)^2$ and the renormalized charge is given by $e_R^2(p^2) = e_R^2(0) \left\{ 1 + \frac{e_R^2(0)}{12\pi^2} \left[\ln \frac{-p^2}{m^2} - \frac{5}{3} - \left(\frac{2c^{pp}}{p^2} \right) \left(\ln \frac{-p^2}{m^2} - \frac{2}{3} \right) \right] \right\}$. There is also a e^2 correction in the vertex, $\Gamma_\mu(p^2) = \gamma_\mu + c_{\alpha\mu}\gamma^\alpha - \frac{e^2}{12\pi^2}c_{\alpha\mu}\gamma^\alpha \left(\ln \frac{-p^2}{m^2} - \frac{5}{3} \right)$.

We run a simulation in order to see how the couplings evolve with these corrections. This is depicted in Figs. 1 and 2 for the QED and QCD coupling, respectively. Bounds on the coefficients for Lorentz violation that come from α measurements are not so stringent; although α is very accurately known at the zero point, its measurements at higher energies have only a precision of two significant figures. However, we know the same thing happens in the running of the strong coupling α_s and QCD observables are affected by this running, so we can see how coefficients change these observables.

The result of the high-energy limit in Eq. (5) can also be used for quarks except for color and flavor factors. In the QCD computation, we also have to consider self-interacting gluon diagrams and the coefficients $k_G^{\mu\nu\alpha\beta}$ and k_{AG}^μ. The former can be taken to be traceless, i.e., $k_{G\ \nu}^{\mu\nu\alpha} = 0$, because a suitable choice of coordinates can absorb this term into $c^{\mu\alpha}$. The latter decreases with the energy scale so its effects are expected to be negligible at high energies.[4] In this way, we can focus on the c-type coefficients.

The Altarelli–Parisi equations depend on the running of α_s. Hence, a perturbative correction of the type c^{pp}/p^2 implies the PDFs depend on Lorentz violation. It is interesting to compare this observation with recent work which shows the leading-twist unpolarized PDF may implicitly depend c^{pp}/Λ_{QCD}^2, but instead through nonperturbative effects. Quantum corrections also affect observables such as the hadronic R ratio, since it also depends on the running of α_s. Here we find

$$R = R_0 \left\{ 1 + \frac{\alpha_s(p^2)}{\pi} + \frac{\alpha_s^2(0)}{\pi} \frac{2c^{pp}}{p^2} n_f \left(\ln \frac{-p^2}{\lambda^2} - \frac{2}{3} \right) + ... \right\}, \qquad (7)$$

where R_0 is the tree level ratio and n_f is the number of flavors.

The QCD sector of the SME also affects the tree-level ratio R_0 besides the running of α_s. There is a huge amount of data available on the R ratio from the Particle Data Group.[6] It is possible to run a simulation in order to constrain sidereal-varying coefficients that appear using these measurements. We adapt the sidereal-time simulation of Refs. 2 for the first one hundred of these R ratio measurements. Note this sidereal simulation is only possible if we include quantum corrections because the tree-level Lorentz-violating correction is not energy-scale dependent as the R ratio measurements are. We present the best limits in Table 1.

Table 1: Constraints on quark sidereal coefficients.

Coefficient	Constraint
$\lvert c_q^{XY} \rvert$	$< 7.9 \times 10^{-2}$
$\lvert c_q^{YZ} \rvert$	$< 3.4 \times 10^{-2}$
$\lvert c_q^{XZ} \rvert$	$< 3.5 \times 10^{-2}$
$\lvert c_q^{XX} - c_q^{YY} \rvert$	$< 1.6 \times 10^{-1}$

References

1. *Data Tables for Lorentz and CPT Violation,* V.A. Kostelecký and N. Russell, 2019 edition, arXiv:0801.0287v12.
2. V.A. Kostelecký , E. Lunghi, and A.R. Vieira, Phys. Lett. B **769**, 272 (2017); E. Lunghi and N. Sherrill, Phys. Rev. D **98**, 115018 (2018); V.A. Kostelecký and Z. Li, Phys. Rev. D **99**, 056016 (2019).
3. A.R. Vieira, A.L. Cherchiglia, and M. Sampaio, Phys. Rev. D **93**, 025029 (2016).
4. D. Colladay and P. McDonald, Phys. Rev. D **75**, 105002 (2007).
5. E. Lunghi, these proceedings; N. Sherrill, these proceedings; V.A. Kostelecký, E. Lunghi, N. Sherrill, and A.R. Vieira, arXiv:1911.04002.
6. M. Tanabashi *et al.*, Phys. Rev. D **98**, 030001 (2018).

Nonminimal Lorentz-Violating Extensions
of Gauge Field Theories

Zonghao Li

Physics Department, Indiana University, Bloomington, IN 47405, USA

A general method is presented to build all gauge-invariant terms in gauge field theories, including quantum electrodynamics and quantum chromodynamics. It is applied to two experiments, light-by-light scattering and deep inelastic scattering, to extract first bounds on certain nonminimal coefficients for Lorentz violation.

1. Introduction

Lorentz violation has been a popular topic in recent years in the search for new physics beyond the Standard Model (SM). The Standard-Model Extension (SME) developed by D. Colladay and V.A. Kostelecký studies Lorentz violation in the context of effective field theory.[1] It includes all possible Lorentz-violating modifications to the SM coupled to General Relativity to describe all possible Lorentz-violating experimental signals. All minimal terms (mass dimensions $d \leq 4$) have been established;[1,2] most nonminimal free-propagation terms have been established;[3] and some low-dimension ($d \leq 6$) interaction terms in quantum electrodynamics (QED) have been established.[4] However, general Lorentz-violating terms in gauge field theories are still unknown. Here, we present a general method to build all Lorentz-violating terms in gauge field theories and apply these to two experiments, light-by-light scattering and deep inelastic scattering (DIS), to get first bounds on certain SME coefficients. Related techniques can be applied in the gravity context.[5] The present contribution to the CPT'19 proceedings is based on results in Ref. 6.

2. Theory

The SME preserves gauge invariance, so we need to find all gauge-invariant terms to build general Lorentz-violating extensions of gauge field theories. The gauge-covariant operator is a powerful tool in building gauge-invariant terms. An operator $\mathcal{O}$ is called gauge covariant if it transforms to $U\mathcal{O}U^\dagger$ under the gauge transformation, where U is a unitary representation of the gauge group $\mathcal{G}$, and the fermion field ψ transforms to $U\psi$ under the gauge transformation. If $\mathcal{O}, \mathcal{O}_1, \mathcal{O}_2$ are gauge-covariant operators, we can build gauge-invariant operators by taking traces of them, $\mathrm{Tr}(\mathcal{O})$, or combining them with Dirac bispinors, $\overline{(\mathcal{O}_1\psi)}(\mathcal{O}_2\psi)$. Therefore, we can first build gauge-covariant operators and then get gauge-invariant operators.

A direct calculation shows that the gauge-covariant derivative D_μ and the gauge field strength tensor $F_{\alpha\beta}$ are gauge-covariant operators. Moreover, we find that any operator formed as a mixture of D and F is gauge covariant. In principle, we can construct gauge-invariant operators from all those operators. However, this would introduce a lot of redundancies because F is related to the commutator of D with itself. Therefore, we need to characterize those gauge-covariant operators in terms of a set of standard bases with controlled or no redundancy. The key result is that any operator formed as a mixture of D and F can be expressed as a linear combination of operators of the form

$$(D_{(n_1)}F_{\beta_1\gamma_1})(D_{(n_2)}F_{\beta_2\gamma_2})\cdots(D_{(n_m)}F_{\beta_m\gamma_m})D_{(n_{m+1})}, \tag{1}$$

where $D_{(n)} = (1/n!)\sum D_{\alpha_1}D_{\alpha_2}\cdots D_{\alpha_n}$ is totally symmetrized with the summation performed over all permutations of $\alpha_1, \alpha_2, \cdots, \alpha_n$. The basic idea behind Eq. (1) is absorbing the symmetric parts in the totally symmetrized $D_{(n)}$ and the antisymmetric parts in F. The detailed proof uses Young tableaux and can be found in Ref. 6.

We proceed to build general gauge-invariant terms from Eq. (1). Both QED and quantum chromodynamics (QCD) are based on gauge field theories, so the general Lorentz-violating extensions of QED and QCD can be constructed. We remark in passing that the extensions include both Lorentz-invariant and Lorentz-violating terms, so we are actually building general gauge-invariant extensions of QED and QCD. The reader is referred to Ref. 6 for the details of the Lagrange densities. In the next two sections, we look at two experimental applications.

3. Light-by-light scattering

Light-by-light scattering is a nonlinear effect of the electromagnetic field, which is hidden in the classical linear Maxwell equations but can arise in QED via radiative loop corrections. Experimental measurements of light-by-light scattering can provide important tests of QED. Since the cross section is tiny, light-by-light scattering was directly measured only recently at the LHC by the ATLAS collaboration.[7] They measured ultraperipheral Pb+Pb collisions at $\sqrt{s_{NN}} = 5.02\,\text{TeV}$. By the equivalent-photon approximation,[8] the collision of high-energy ultraperipheral heavy ions can be treated as collisions of photons from the heavy ions.

The QED extension built in the last section can describe all possible deviations from the SM prediction in light-by-light scattering experiments. The dominant contribution comes from a $d = 8$ term:

$$\mathcal{L}_g^{(8)} \supset -\tfrac{1}{48} k_F^{(8)\kappa\lambda\mu\nu\rho\sigma\tau\upsilon} F_{\kappa\lambda} F_{\mu\nu} F_{\rho\sigma} F_{\tau\upsilon}, \tag{2}$$

where $k_F^{(8)\kappa\lambda\mu\nu\rho\sigma\tau\upsilon}$ are coefficients for Lorentz violation. This term creates a new interaction vertex with four photon lines and contributes to the light-by-light scattering at tree level. Many possible Lorentz-violating signals can arise from this. It produces new contributions to the total cross section of light-by-light scattering in addition to the SM ones. The SME coefficients are assumed to be approximately constant in the Sun-centered frame,[9] so the experimental cross section can depend on the sidereal time with the Earth rotating about its axis and revolving around the Sun. The experimental results can also depend on the location and orientation of the laboratory. The Lorentz-violating term can produce a new energy dependence for the differential cross section as well.

Due to statistical limitations of the data, we compare here only the total cross sections to get bounds on the SME coefficients. Future improvements in the experiment can lead to more detailed investigations of possible Lorentz-violating signals. The LHC experiment measured the total cross section as $70 \pm 24(\text{stat.}) \pm 17(\text{syst.})\,\text{nb}$.[7] The theoretical SM prediction is $49 \pm 10\,\text{nb}$.[10] Comparing these two results gives bounds[6] on 126 components of the coefficients $k_F^{(8)}$. The bounds on the Lorentz-invariant and isotropic components of the coefficients $k_F^{(8)}$ are also extracted. All these components are constrained to approximately $10^{-7}\,\text{GeV}^{-4}$.

4. Deep inelastic scattering

DIS provides key experimental support for the existence of quarks and the predictions of QCD. It is also an essential tool in the search for new physics beyond the SM and can be used to test Lorentz symmetry. The QCD+QED extension built via the method presented in Sec. 2 can describe all Lorentz-violating signals in DIS experiments.

The contributions from the minimal SME to DIS have been considered before.[11] Here, we focus on contributions from nonminimal terms.

Since most DIS experiments use unpolarized beams, we consider spin-independent terms. The leading-order spin-independent contribution from the nonminimal SME is

$$\mathcal{L}_\psi^{(5)} \supset -\tfrac{1}{2} a_f^{(5)\mu\alpha\beta} \overline{\psi}_f \gamma_\mu i D_{(\alpha} i D_{\beta)} \psi_f + \text{h.c.}, \tag{3}$$

where the parentheses around the lower indices mean symmetrization on α and β with a factor of $1/2$, $f = u, d$ includes the dominant quark flavors, and $a_f^{(5)\mu\alpha\beta}$ are $d = 5$ coefficients for Lorentz violation. This term modifies both the free propagation of fermions and interactions among photons and fermions. The cross section with the corrections can be found in Ref. 6.

Based on the simulations in Ref. 11 for $c^{\mu\nu}$ coefficients, we can estimate the bounds on $a^{(5)\mu\alpha\beta}$ to be around $10^{-7} - 10^{-4}\,\text{GeV}^{-1}$. We can also expect that the corrected DIS cross section depends on up to the third-order harmonics of the sidereal-time variables because the coefficients $a^{(5)\mu\alpha\beta}$ contain three indices. The $a^{(5)\mu\alpha\beta}$ coefficients also provide CPT-odd contributions to DIS for protons and antiprotons. Experimental measurements of those Lorentz-violating signals can provide fruitful insight into new physics beyond the SM.

Acknowledgments

This work was supported in part by the US Department of Energy and by the Indiana University Center for Spacetime Symmetries.

References

1. D. Colladay and V.A. Kostelecký, Phys. Rev. D **55**, 6760 (1997); Phys. Rev. D **58**, 116002 (1998).
2. V.A. Kostelecký, Phys. Rev. D **69**, 105009 (2004).
3. V.A. Kostelecký and M. Mewes, Phys. Rev. D **80**, 015020 (2009); Phys. Rev. D **88**, 096006 (2013); Phys. Rev. D **85**, 096005 (2012); Phys. Lett. B **779**, 136 (2018).
4. Y. Ding and V.A. Kostelecký, Phys. Rev. D **94**, 056008 (2016).
5. V.A. Kostelecký and Z. Li, in preparation.
6. V.A. Kostelecký and Z. Li, Phys. Rev. D **99**, 056016 (2019).
7. M. Aaboud *et al.*, Nat. Phys. **13**, 852 (2017).
8. C. von Weizsäcker, Z. Phys. **88**, 612 (1934); E.J. Williams, Phys. Rev. **45**, 729 (1934); E. Fermi, Nuov. Cim. **2**, 143 (1925).
9. *Data Tables for Lorentz and CPT Violation*, V.A. Kostelecký and N. Russell, 2019 edition, arXiv:0801.0287v12.
10. M. Kłusek-Gawenda and A. Szczurek, Phys. Rev. C **82**, 014904 (2010).
11. V.A. Kostelecký, E. Lunghi, and A.R. Vieira, Phys. Lett. B **769**, 272 (2017).

Constraining Dimension-Six Nonminimal Lorentz-Violating Electron–Nucleon Interactions with EDM Physics

Jonas B. Araujo,* A.H. Blin,[†] Marcos Sampaio,[‡] and Manoel M. Ferreira Jr*

*Departamento de Física, Universidade Federal do Maranhão,
Campus Universitário do Bacanga, São Luís - MA, 65080-805, Brazil

[†] CFisUC, Department of Physics, University of Coimbra,
3004-516 Coimbra, Portugal

[‡] CCNH, Universidade Federal do ABC, Santo André - SP, 09210-580, Brazil

Electric dipole moments of atoms can arise from P-odd and T-odd electron–nucleon couplings. This work studies a general class of dimension-six electron–nucleon interactions mediated by Lorentz-violating tensors of ranks ranging from 1 to 4. The possible couplings are listed as well as their behavior under C, P, and T, allowing us to select the couplings compatible with electric-dipole-moment physics. The unsuppressed contributions of these couplings to the atom's hamiltonian can be read as equivalent to an electric dipole moment. The Lorentz-violating coefficients' magnitudes are limited using electric-dipole-moment measurements at the levels of $3.2 \times 10^{-31} (\text{eV})^{-2}$ or $1.6 \times 10^{-33} (\text{eV})^{-2}$.

1. Introduction

Electric dipole moments (EDMs) are excellent probes for violations of discrete symmetries and for physics beyond the Standard Model.[1] EDM terms violate both parity (P) and time-reversal (T) symmetry, while preserving charge conjugation (C), assuming the CPT theorem holds. An atom's EDM could arise from intrinsic properties of the electrons and/or nucleus, or from P- and T-odd electron–nucleon (e–N) couplings. EDM phenomenology can also arise in a Lorentz-violation (LV) scenario addressed within the framework of the Standard-Model Extension (SME).[2] LV generates CP violation and EDMs via radiative corrections[3] or at tree level via dimension-five nonminimal couplings,[4,5] and may also yield nuclear-EDM corrections to the Schiff moment.[6] Nonminimal couplings have been of great interest in recent years.[7] In this work, we investigate a class of dimension-six LV e–N couplings, composed of rank-1 to rank-4 background tensors, first proposed in Ref. 8, and the possible generation of atomic EDMs.[9]

2. Nonminimal e–N Lorentz-violating couplings

The simplest couplings involve a rank-1 LV tensor $(k_{XX})_\mu$ with an effective lagrangian of the form $\mathcal{L}_{\text{LV}} = (k_{XX})_\mu \left[(\bar{N} \, \Gamma_1 \, N) \left(\bar{\psi} \, \Gamma_2 \, \psi \right) \right]^\mu$, where the upper μ belongs either to Γ_1 or Γ_2. The subscript XX labels the type of fermion bilinear: scalar (S), pseudoscalar (P), vector (V), axial vector (A), and tensor (T), accounting for the 16 linearly independent 4×4 matrices. The operators $\Gamma_{1,2}$ must be combinations of Dirac matrices, given here as

$$\alpha^i = \begin{pmatrix} 0 & \sigma^i \\ \sigma^i & 0 \end{pmatrix}, \ \Sigma^k = \begin{pmatrix} \sigma^k & 0 \\ 0 & \sigma^k \end{pmatrix}, \ \gamma^0 = \begin{pmatrix} 1 & 0 \\ 0 & -1 \end{pmatrix}, \ \gamma^5 = \begin{pmatrix} 0 & 1 \\ 1 & 0 \end{pmatrix}, \quad (1)$$

and $\gamma^i = \gamma^0 \alpha^i$, $\sigma^{\mu\nu} = i \left[\gamma^\mu, \gamma^\nu \right] / 2$, with $\sigma^{0j} = i\alpha^j$, $\sigma^{ij} = \epsilon_{ijk} \Sigma^k$. As we are interested in EDM behavior, we select the P-odd and T-odd components. The rank-1 couplings are listed in Table 1, in which "S" and "NS" mean "suppressed" and "not suppressed," respectively, in the nucleon's nonrelativistic limit. Couplings of higher ranks are shown in Tables 2 to 4.

Table 1: General CPT-odd couplings with a rank-1 LV tensor.

Coupling	P-Odd, T-Odd Piece	NRL	EDM
$(k_{SV})_\mu \, (\bar{N}N)(\bar{\psi}\gamma^\mu\psi)$	$(k_{SV})_i \, (\bar{N}N)(\bar{\psi}\gamma^i\psi)$	NS	yes
$(k_{VS})_\mu \, (\bar{N}\gamma^\mu N)(\bar{\psi}\psi)$	$(k_{VS})_i \, (\bar{N}\gamma^i N)(\bar{\psi}\psi)$	S	–
$(k_{VP})_\mu \, (\bar{N}\gamma^\mu N)(\bar{\psi}i\gamma_5\psi)$	$(k_{VP})_0 \, (\bar{N}\gamma^0 N)(\bar{\psi}i\gamma_5\psi)$	NS	yes
$(k_{PV})_\mu \, (\bar{N}i\gamma_5 N)(\bar{\psi}\gamma^\mu\psi)$	$(k_{PV})_0 \, (\bar{N}\gamma_5 N)(\bar{\psi}\gamma^0\psi)$	S	–
$(k_{SA})_\mu \, (\bar{N}N)(\bar{\psi}\gamma^\mu\gamma_5\psi)$	none	–	–
$(k_{AS})_\mu \, (\bar{N}\gamma^\mu\gamma_5 N)(\bar{\psi}\psi)$	none	–	–
$(k_{PA})_\mu \, (\bar{N}i\gamma_5 N)(\bar{\psi}\gamma^\mu\gamma_5\psi)$	none	–	–
$(k_{AP})_\mu \, (\bar{N}\gamma^\mu\gamma_5 N)(\bar{\psi}i\gamma_5\psi)$	none	–	–

The EDM contribution, in the nonrelativistic limit for the nucleons, is calculated via atomic parity nonconservation methods[10] as

$$\frac{\Delta E}{E_z} = 2e\text{Re}\left[\langle\psi_0|H_{P,T}|\eta\rangle\right] \equiv d_{\text{equiv}}, \quad (2)$$

where $H_{P,T}$ is the coupling's hamiltonian contribution for the electron, and $|\eta\rangle, |\psi_0\rangle$—which have opposite parity—correspond to

$$(\psi_0)^l = \begin{pmatrix} \frac{i}{r} G_{l,J=\frac{1}{2}}(r)\phi^l_{\frac{1}{2},\frac{1}{2}} \\ \frac{1}{r} F_{l,J=\frac{1}{2}}(r)\boldsymbol{\sigma}\cdot\hat{r}\phi^l_{\frac{1}{2},\frac{1}{2}} \end{pmatrix}, \ \eta^l = \begin{pmatrix} \frac{i}{r} G^S_{l,J=\frac{1}{2}}(r)\phi^l_{\frac{1}{2},\frac{1}{2}} \\ \frac{1}{r} F^S_{l,J=\frac{1}{2}}(r)\boldsymbol{\sigma}\cdot\hat{r}\phi^l_{\frac{1}{2},\frac{1}{2}} \end{pmatrix}, \quad (3)$$

where $\phi^{l=0}_{\frac{1}{2},\frac{1}{2}} = (Y^0_0, 0)^T$ and $\phi^{l=1}_{\frac{1}{2},\frac{1}{2}} = (\sqrt{\frac{1}{3}}Y^0_1, -\sqrt{\frac{2}{3}}Y^1_1)^T$. While $(\psi_0)^l$ obeys Dirac's equation for a central potential, η^l obeys Sternheimer's equation.[10]

Table 2: General CPT-even couplings with a rank-2 LV tensor.

Coupling	P-Odd, T-Odd Piece	NRL	EDM
$(k_{VV})_{\mu\nu} \left(\bar N\gamma^\mu N\right) \left(\bar\psi\gamma^\nu\psi\right)$	$(k_{VV})_{i0} \left(\bar N\gamma^i N\right) \left(\bar\psi\gamma^0\psi\right)$	S	–
	$(k_{VV})_{0i} \left(\bar N\gamma^0 N\right) \left(\bar\psi\gamma^i\psi\right)$	NS	yes
$(k_{AV})_{\mu\nu} \left(\bar N\gamma^\mu\gamma_5 N\right) \left(\bar\psi\gamma^\nu\psi\right)$	none	–	–
$(k_{VA})_{\mu\nu} \left(\bar N\gamma^\mu N\right) \left(\bar\psi\gamma^\nu\gamma_5\psi\right)$	none	–	–
$(k_{AA})_{\mu\nu} \left(\bar N\gamma^\mu\gamma_5 N\right) \left(\bar\psi\gamma^\nu\gamma_5\psi\right)$	$(k_{AA})_{0i} \left(\bar N\gamma^0\gamma_5 N\right) \left(\bar\psi\gamma^i\gamma_5\psi\right)$	S	–
	$(k_{AA})_{i0} \left(\bar N\gamma^i\gamma_5 N\right) \left(\bar\psi\gamma^0\gamma_5\psi\right)$	NS	yes
$(k_{TS})_{\mu\nu} \left(\bar N\sigma^{\mu\nu} N\right) \left(\bar\psi\psi\right)$	none	–	–
$(k_{TP})_{\mu\nu} \left(\bar N\sigma^{\mu\nu} N\right) \left(\bar\psi i\gamma_5\psi\right)$	none	–	–
$(k_{ST})_{\mu\nu} \left(\bar N N\right) \left(\bar\psi\sigma^{\mu\nu}\psi\right)$	none	–	–
$(k_{PT})_{\mu\nu} \left(\bar N i\gamma_5 N\right) \left(\bar\psi\sigma^{\mu\nu}\psi\right)$	none	–	–

Table 3: General CPT-odd couplings with a rank-3 LV tensor.

Coupling	P-Odd, T-Odd Piece	NRL	EDM
$(k_{VT})_{\alpha\mu\nu} \left(\bar N\gamma^\alpha N\right)(\bar\psi\sigma^{\mu\nu}\psi)$	none	–	–
$(k_{AT})_{\alpha\mu\nu} \left(\bar N\gamma^\alpha\gamma_5 N\right)(\bar\psi\sigma^{\mu\nu}\psi)$	$(k_{AT})_{0ij} \left(\bar N\gamma^0\gamma_5 N\right)(\bar\psi\sigma^{ij}\psi)$	S	–
	$(k_{AT})_{i0j} \left(\bar N\gamma^i\gamma_5 N\right)(\bar\psi\sigma^{0j}\psi)$	NS	yes
	$(k_{AT})_{ij0} \left(\bar N\gamma^i\gamma_5 N\right)(\bar\psi\sigma^{j0}\psi)$	NS	yes
$(k_{TV})_{\alpha\mu\nu} \left(\bar N\sigma^{\mu\nu} N\right)(\bar\psi\gamma^\alpha\psi)$	none	–	–
$(k_{TA})_{\alpha\mu\nu} \left(\bar N\sigma^{\mu\nu} N\right)(\bar\psi\gamma^\alpha\gamma_5\psi)$	$(k_{TA})_{0ij} \left(\bar N\sigma^{ij} N\right)(\bar\psi\gamma^0\gamma_5\psi)$	NS	yes
	$(k_{TA})_{i0j} \left(\bar N\sigma^{0j} N\right)(\bar\psi\gamma^i\gamma_5\psi)$	S	–
	$(k_{TA})_{ij0} \left(\bar N\sigma^{j0} N\right)(\bar\psi\gamma^i\gamma_5\psi)$	S	–

Table 4: General CPT-even couplings with a rank-4 LV tensor.

Coupling	P-Odd, T-Odd Piece	NRL	EDM
$(k_{TT})_{\alpha\beta\mu\nu} \left(\bar N\sigma^{\alpha\beta} N\right)(\bar\psi\sigma^{\mu\nu}\psi)$	$(k_{TT})_{0ijk} \left(\bar N\sigma^{0i} N\right)(\bar\psi\sigma^{jk}\psi)$	S	–
	$(k_{TT})_{i0jk} \left(\bar N\sigma^{i0} N\right)(\bar\psi\sigma^{jk}\psi)$	S	–
	$(k_{TT})_{ij0k} \left(\bar N\sigma^{ij} N\right)(\bar\psi\sigma^{0k}\psi)$	NS	yes
	$(k_{TT})_{ijk0} \left(\bar N\sigma^{ij} N\right)(\bar\psi\sigma^{k0}\psi)$	NS	yes

By evaluating d_{equiv} for each coupling via Eq. (2) and performing a sidereal analysis,[5] one can set upper bounds on the LV coefficients using numerical estimates on the thallium atom and the experimental limit on the electron's EDM.[11] A list of time-averaged upper bounds is given in Table 5. More information can be found in Ref. 12.

Table 5: List of bounds on the LV tensors of ranks ranging from 1 to 4.

Component	Upper Bound
$\lvert (k_{VP})^{(\text{Sun})}_0 \rvert$	$1.6 \times 10^{-33} (\text{eV})^{-2}$
$\lvert \frac{1}{4}\left[(k_{AT})^{(\text{Sun})}_{101} + (k_{AT})^{(\text{Sun})}_{202} - 2\,(k_{AT})^{(\text{Sun})}_{303} \right] \sin 2\chi \rvert$	$3.2 \times 10^{-31} (\text{eV})^{-2}$
$\lvert \left[-(k_{AT})^{(\text{Sun})}_{102} + (k_{AT})^{(\text{Sun})}_{201} \right] \sin \chi \rvert$	$3.2 \times 10^{-31} (\text{eV})^{-2}$
$\lvert \frac{1}{2}\left((k_{AT})^{(\text{Sun})}_{101} + (k_{AT})^{(\text{Sun})}_{202} \right) \sin^2 \chi + (k_{AT})^{(\text{Sun})}_{303} \cos^2 \chi \rvert$	$3.2 \times 10^{-31} (\text{eV})^{-2}$
$\lvert \frac{1}{4}\left[(K_{TT})^{(\text{Sun})}_{011} + (K_{TT})^{(\text{Sun})}_{022} - 2\,(K_{TT})^{(\text{Sun})}_{033} \right] \sin 2\chi \rvert$	$3.2 \times 10^{-31} (\text{eV})^{-2}$
$\lvert \left[(K_{TT})^{(\text{Sun})}_{012} - (K_{TT})^{(\text{Sun})}_{021} \right] \sin \chi \rvert$	$3.2 \times 10^{-31} (\text{eV})^{-2}$
$\lvert \frac{1}{2}\left((K_{TT})^{(\text{Sun})}_{011} + (K_{TT})^{(\text{Sun})}_{022} \right) \sin^2 \chi + (K_{TT})^{(\text{Sun})}_{033} \cos^2 \chi \rvert$	$3.2 \times 10^{-31} (\text{eV})^{-2}$

Acknowledgments

We are grateful to CAPES, CNPq, FCT Portugal, and FAPEMA.

References

1. J. Engel, M.J. Ramsey-Musolf, and U. van Kolck, Prog. Part. Nucl. Phys. **71**, 21 (2013); T. Chupp and M.J. Ramsey-Musolf, Phys. Rev. C **91**, 035502 (2015); N. Yamanaka *et al.*, Eur. Phys. J. A **53**, 54 (2017).

2. D. Colladay and V.A. Kostelecký, Phys. Rev. D **55**, 6760 (1997); Phys. Rev. D **58**, 116002 (1998); S.R. Coleman and S.L. Glashow, Phys. Rev. D **59**, 116008 (1999).

3. M. Haghighat, I. Motie, and Z. Rezaei, Int. J. Mod. Phys. A **28**, 1350115 (2013).

4. P.A. Bolokhov, M. Pospelov, and M. Romalis, Phys. Rev. D **78**, 057702 (2008).

5. J.B. Araujo, R. Casana, and M.M. Ferreira, Phys. Rev. D **92**, 025049 (2015); Phys. Lett. B **760**, 302 (2016).

6. J.B. Araujo, R. Casana, and M.M. Ferreira, Phys. Rev. D **97**, 055032 (2018).

7. G. Gazzola *et al.*, J. Phys. G **39**, 035002 (2012); L.C.T. Brito, H.G. Fargnoli, and A.P. Bata Scarpelli, Phys. Rev. D **87**, no. 12, 125023 (2013); L.H.C. Borges *et al.*, Phys. Lett. B **756**, 332 (2016); Y.M.P. Gomes and J.T. Guaitolini Junior, Phys. Rev. D **99**, 055006 (2019); V.E. Mouchrek-Santos and M.M. Ferreira, Phys. Rev. D **95**, no. 7, 071701 (2017); J. Phys. Conf. Ser. **952**, 012019 (2018).

8. V.A. Kostelecký and Z. Li, Phys. Rev. D **99**, 056016 (2019).

9. N. Yamanaka, Int. J. Mod. Phys. E **26**, 1730002 (2017).

10. B.L. Roberts and W.J. Marciano, Adv. Ser. Direct. High Energy Phys. **20**, 1 (2009).

11. ACME Collaboration, V. Andreev *et al.*, Nature **562**, 355 (2018).

12. J.B. Araujo *et al.*, arXiv:1902.10329, submitted for publication in Phys. Rev. D.

First Demonstration of Antimatter Quantum Interferometry

M. Giammarchi

Istituto Nazionale di Fisica Nucleare, Sezione di Milano,
Via Celoria 16, 20133, Italy

On behalf of the QUPLAS Collaboration

This paper describes the first experimental evidence of antimatter-wave interference, a process at the heart of quantum physics and its interpretation. For the case of ordinary matter particles, interference phenomena have been observed in a variety of cases, ranging from electrons up to complex molecules. Here, I present the first demonstration of single-positron quantum interference.

1. Introduction

The concept of wave–particle duality was introduced in 1923 by de Broglie:[1] the Planck constant h relates the momentum of a massive particle p to a wavelength $\lambda_{dB} = h/p$. Accordingly, diffraction and interference phenomena have been observed for a variety of particles, ranging from electrons,[2,3] to neutrons[4,5] to complex molecules.[6–8]

The experimental study of single-particle (one-at-a-time) double-slit interference was proposed by Feynman as a decisive test that *has in it the heart of quantum mechanics.*[9] A single-electron interference experiment was conducted for the first time in 1976.[10] A demonstration of single-particle interference for any antimatter particle, however, has been missing.

In order to fill this gap, we have realized a Talbot–Lau interferometer[11] suitable for anti-electrons (positrons) in the 5–18 keV energy range. This development is part of the QUPLAS (QUantum interferometry and gravitation with Positrons and LASers) research program.[12–15]

2. The setup

The experiment is located at the variable-energy positron-beam facility of L-NESS[a] at the Politecnico di Milano in Como (Italy). The positron beam

---

[a]Laboratory for Nanostructure Epitaxy and Spintronics on Silicon

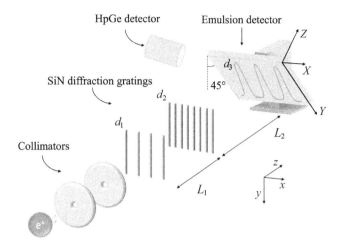

Fig. 1: Sketch of the setup: the beam, prepared by two 2-mm-wide collimators, reaches the gratings and the emulsion detector. In the setup, $L_1 = 11.8$ cm and $L_2 = 57.6$ cm. The emulsion is tilted by 45 degrees in order to scan better the longitudinal variation of visible fringes. An HpGe detector is used to monitor the beam flux.

has an intensity of $\sim 5 \times 10^3 \, e^+/$s, an energy between 5 and 18 keV (resolution better than 0.1%) and an angular divergence of a few milliradians producing a spot of about 2 mm. The beam passes through the interferometer and the hits the emulsion detector as shown in Fig. 1. The interferometer consists of two gratings in a period-magnifying Talbot–Lau configuration.[13] The two gold-coated SiN gratings (11.8 cm apart) have nominal periods of $d_1 = 1.2 \, \mu$m and $d_2 = 1 \, \mu$m, respectively, and a 50% open fraction producing $d_3 = 5.9 \, \mu$m period fringes at the location of the emulsion detector.

The periodic spatial distribution is recorded by the emulsion detector, which has a $\sim \mu$m resolution that has been tested at QUPLAS energies in a dedicated "engineering" run.[15] The setup, aligned as described in Ref. 16, features a moderate flux ($\sim 10^3/$s) produced by an incoherent (radioactive) source. Since the transit time of the particles in the interferometer is of about 10^{-8} s, QUPLAS is a single-particle experiment.

3. Results

The configuration of the interferometer was meeting the resonant condition for the nominal 14 keV energy value, as[13]

$$\frac{L_1}{L_2} = \frac{d_1}{d_2} - 1. \tag{1}$$

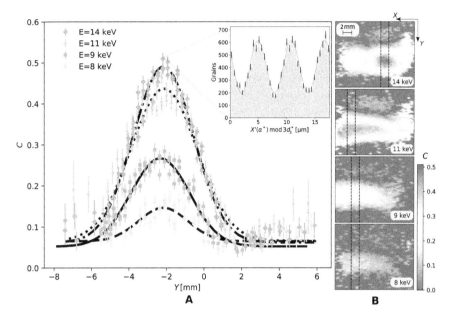

Fig. 2: Panel A: Contrast C as a function of the longitudinal coordinate y for the different energies in the regions delimited by the dashed lines in Panel B. The insert shows the actual periodicity for the case of best visibility. Panel B: Contrast as a function of position on the emulsion for the different energies shown.

Exposures of the emulsions were made at 8, 9, 11, 14, and 16 keV positron energies; the results are presented in Fig. 2. For each energy, the emulsion was exposed for about 100 hours, then developed, and the impact positions of positrons were digitized. The analysis strategy was to fit the fringe distribution as a function of both the period and the rotation angle between the interferometer and the emulsion detector (the remaining important alignment parameter). Periodical patterns, as the one shown in the insert of Fig. 2, have been obtained with the expected periodicity of $5.8\,\mu$m.

After having shown periodical patterns as expected, in order to fully prove the quantum nature of the effect, the visibility was studied as a function of the energy (wavelength). As shown in Fig. 3, the fringe contrast as a function of energy disagrees with projective classical mechanics and is in agreement with the quantum-mechanical model of the system.[12,13] This is the first demonstration of antimatter quantum interference.

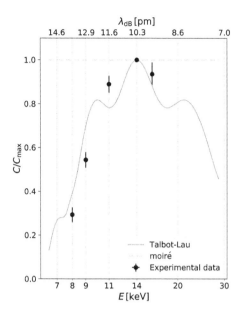

Fig. 3: The visibility as a function of energy proves the quantum-mechanical nature of the effect. Its classical (Moirè) counterpart would have been wavelength independent.

References

1. L. de Broglie, Nature **112**, 540 (1923).
2. C.J. Davisson and L.H. Germer, Proc. Natl. Acad. Sci. USA **14**, 317 (1928).
3. G.P. Thomson and A. Reid, Nature **119**, 890 (1927).
4. H. Rauch *et al.*, Phys. Lett. A **47**, 369 (1974).
5. A. Zeilinger *et al.*, Rev. Mod. Phys. **60**, 1067 (1988).
6. M.S. Chapman *et al.*, Phys. Rev. Lett. **74**, 4783 (1995).
7. M. Arndt *et al.*, Nature **401**, 680 (1999).
8. B. Brezger *et al.*, Phys. Rev. Lett. **88**, 100404 (2002).
9. R.P. Feynman, R.B. Leighton, and M. Sands, *Feynman Lectures on Physics*, Vol. 3, Addison-Wesley, Boston, 1965.
10. P.G. Merli *et al.*, Am. J. Phys. **44**, 306 (1976).
11. J.F. Clauser and S. Li, Phys. Rev. A **49**, R2213 (1994).
12. S. Sala *et al.*, J. Phys. B **48**, 195002 (2015).
13. S. Sala, M. Giammarchi, and S. Olivares, Phys. Rev. A **94**, 033625 (2016).
14. S. Aghion *et al.*, JINST **11**, P06017 (2016).
15. S. Aghion *et al.*, JINST **13**, P05013 (2018).
16. S. Sala *et al.*, Sci. Adv. **5**, EAAV7610 (2019).

114

Cryogenic Penning-Trap Apparatus for Precision Experiments with Sympathetically Cooled (Anti)Protons

M. Niemann,* T. Meiners,* J. Mielke,* N. Pulido,* J. Schaper,¶,* M.J. Borchert,¶,*
J.M. Cornejo,* A.-G. Paschke,*,† G. Zarantonello,*,† H. Hahn,†,* T. Lang,*
C. Manzoni,‡ M. Marangoni,‡ G. Cerullo,‡ U. Morgner,* J.-A. Fenske,†
A. Bautista-Salvador,†,* R. Lehnert,§,* S. Ulmer,¶ and C. Ospelkaus*,†

*Institut für Quantenoptik, Leibniz Universität Hannover,
Welfengarten 1, 30167 Hannover, Germany, and
Laboratorium für Nano- und Quantenengineering, Leibniz Universität Hannover,
Schneiderberg 39, 30167 Hannover, Germany

†Physikalisch-Technische Bundesanstalt
Bundesallee 100, 38116 Braunschweig, Germany

‡IFN-CNR, Dipartimento di Fisica, Politecnico di Milano, Piazza L. da Vinci 32,
Milano, 20133, Italy

§Indiana University Center for Spacetime Symmetries Bloomington, IN 47405, USA

¶Ulmer Fundamental Symmetries Laboratory, RIKEN
Hirosawa, Wako, Saitama 351-0198, Japan

Current precision experiments with single (anti)protons to test CPT symmetry progress at a rapid pace, but are complicated by the need to cool particles to sub-thermal energies. We describe a cryogenic Penning-trap setup for $^9Be^+$ ions designed to allow coupling of single (anti)protons to laser-cooled atomic ions for sympathetic cooling and quantum logic spectroscopy. We report on trapping and laser cooling of clouds and single $^9Be^+$ ions. We discuss prospects for a microfabricated trap to allow coupling of single (anti)protons to laser-cooled $^9Be^+$ ions for sympathetic laser cooling to sub-mK temperatures on ms time scales.

1. Introduction

As a result of CPT symmetry, particles and their antiparticles must have the same mass, lifetime, charge, and magnetic moment. The (anti)proton is an attractive candidate to test CPT symmetry by comparing matter–antimatter conjugates in the baryonic sector, complementary to tests in the lepton sector, e.g., with electrons and positrons.[1] Magnetic-moment comparisons[2,3] in particular provide a sensitive test for potential new physics.[4]

In Penning-trap precision measurements, magnetic moments can be determined by measuring the ratio of the Larmor frequency to the free cyclotron frequency. Compared to the electron and positron, the bigger mass of the (anti)proton makes resistive cooling to the ground state of the cyclotron motion impossible. The resulting finite temperature complicates the measurement of the Larmor frequency via the continuous Stern–Gerlach effect.[5] Furthermore, systematic effects can be proportional to the oscillation amplitude of the particle making efficient cooling to sub-thermal energies highly desirable. Such temperatures could be reached through sympathetic cooling with a laser-cooled ion. Implementing this in the same potential well could introduce additional systematic effects and is not possible for the commonly used positively charged laser-cooled ions together with the antiproton. We pursue an approach where the (anti)proton and the laser-cooled ion are confined in spatially separate potential wells and interact remotely via the Coulomb interaction.[6] This approach has already been demonstrated with pairs of atomic ions in radio-frequency Paul traps.[7,8]

2. Trapping single ^{9}Be$^+$ ions

We have built and commissioned a cryogenic Penning-trap system to implement this approach. The setup is based on the BASE CERN setup,[9] enhanced to allow for laser access for ablation loading, photo-ionization, Doppler and ground-state cooling of ^{9}Be$^+$ ions, as well as for detection of laser-induced fluorescence. The setup is operated at a magnetic field strength of 5 T and cooled using a vibration-isolated cryocooler. It is described in more detail in Ref. 10. We load clouds of ^{9}Be$^+$ ions by focusing single 532 nm laser pulses with tens of μJ pulse energy onto a solid beryllium target embedded into a trap electrode via an in-vacuum off-axis parabolic mirror. We laser cool the ions and detect their presence through laser-induced fluorescence using a laser beam at 313 nm. We apply a series of waveforms to the trap electrodes in order to subsequently reduce the number of particles, until single ions can be observed as identified by discrete steps in the level of laser-induced fluorescence.

3. Experimental prospects for ^{9}Be$^+$ ions in Penning traps

We have recently demonstrated spin-motional control of ^{9}Be$^+$ ions using a spectrally tailored frequency comb.[11] While the experimental demonstration was carried out at a comparatively low magnetic field in a microfabricated Paul trap, this type of manipulation lends itself to ground-state cooling of ^{9}Be$^+$ ions in the $5\,$T field of the Penning trap, where electron-spin energy levels with a level spacing of $\sim 140\,$GHz need to be coupled using a stimulated Raman process. Our immediate plans for the setup comprise the demonstration of ground-state cooling and the coupling of pairs of ^{9}Be$^+$ ions as a precursor for sympathetic cooling of (anti)protons.

4. (Anti)protons and quantum logic spectroscopy

We have designed an off-axis proton source for the apparatus, which is awaiting commissioning. Coupling a single (anti)proton to a laser-cooled ^{9}Be$^+$ ion will likely require a microfabricated trap in order to minimize the distance between the particles and obtain a strong coupling. Compared to the commonly used microfabricated surface-electrode Paul traps, a ring-shaped electrode of a micro Penning trap would have to be metal coated both on the front and back face of the disc and on the inside of the ring. We have produced test structures using deep reactive ion etching of a silicon wafer. For metalization, wafers have been coated either using resistive evaporation of Ti and Au under constant rotation of the sample or using direct sputter deposition of Au. On top of these thin metal films, a thick electroplated film of Au has been grown and structured using optical lithography. A sample structure is shown in Fig. 1. The process will be extended by including the spacers to electrically isolate multiple such rings already at the stage of the wafer etching. Electrical isolation will be possible through the use of the lithographic step, which allows us to leave parts of the sample uncoated. Once all of the above ingredients have been implemented and a suitable spin-motional coupling mechanism for the (anti)proton has been implemented, quantum logic spectroscopy[6,12] could be envisioned as a means of probing all relevant transitions of single (anti)protons out of the motional ground state.

Acknowledgments

We acknowledge the support of Wissenschaftlicher Gerätebau at PTB for the construction of the setup and the rotation holder, the support of the PTB cleanroom team concerning the fabrication of the sample structures, and P. Hinze for the SEM pictures. We acknowledge funding from ERC StG "QLEDS," and from DFG through SFB 1227 DQ-*mat* (project B06), DFG grant "Apparatur für kryogene Ionenfallen," and the clusters of excellence QUEST and Quantum Frontiers. RL acknowledges support from the Alexander von Humboldt Foundation.

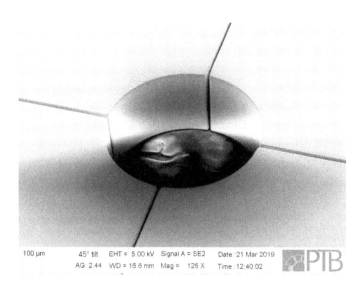

Fig. 1: Gold electroplated silicon test structure for a cylindrical micro Penning trap.

References

1. H. Dehmelt, Rev. Mod. Phys. **62**, 525 (1990).
2. G. Schneider *et al.*, Science **358**, 1081 (2017).
3. C. Smorra *et al.*, Nature **550**, 371 (2017).
4. R. Bluhm, V.A. Kosteleck, and N. Russell, Phys. Rev. Lett. **79**, 1432 (1997).
5. A. Mooser *et al.*, Phys. Rev. Lett. **110**, 140405 (2013).
6. D.J. Wineland *et al.*, J. Res. Natl. Inst. Stand. Tech. **103**, 259 (1998).
7. K.R. Brown *et al.*, Nature **471**, 196 (2011).
8. M. Harlander *et al.*, Nature **471**, 200 (2011).
9. C. Smorra *et al.*, Eur. Phys. J. ST **224**, 3055 (2015).
10. M. Niemann *et al.*, arXiv:1906.09249.
11. A.-G. Paschke *et al.*, Phys. Rev. Lett. **122**, 123606 (2019).
12. D.J. Heinzen and D.J. Wineland, Phys. Rev. A **42**, 2977 (1990).

Tests of Lorentz-Invariance Violation in the Standard-Model Extension with Ultracold Neutrons in qBounce Experiments

A.N. Ivanov,* M. Wellenzohn,*,† and H. Abele*

*Atominstitut, Technische Universität Wien, Stadionallee 2, 1020 Wien, Austria

†FH Campus Wien, University of Applied Sciences,
Favoritenstraße 226, 1100 Wien, Austria

On behalf of the qBounce Collaboration

We analyze corrections, caused by interactions violating Lorentz invariance within the Standard-Model Extension, to the transition frequencies of transitions between quantum gravitational states of unpolarized and polarized ultracold neutrons. Using the current experimental sensitivity of the qBounce experiments $\Delta E < 2 \times 10^{-15}$ eV we make some estimates of upper bounds of parameters for Lorentz-invariance violation in the neutron sector of the Standard-Model Extension that can serve as a theoretical basis for an experimental analysis.

1. Introduction

Low-energy observable effects that may be associated with the breaking of Lorentz symmetry are described by the effective quantum field theory called the Standard-Model Extension (SME).[1] For the experimental analysis of effects predicted by the SME, we propose to use the method of Gravity Resonance Spectroscopy (or the GRS method),[2] which is based on the method of Resonance Spectroscopy or Rabi's method. The GRS method, dealing with quantum gravitational states of ultracold neutrons (UCNs) bouncing in the gravitational field of the Earth, allows the measurement of transition frequencies between quantum gravitational states of UCNs with a current energy resolution (or sensitivity) $\Delta E < 2 \times 10^{-15}$ eV or even better,[3] which is closely related to experimental uncertainties in the measurement.[4] The GRS method has been successfully applied to experimental searches of a large variety of gravitational effects.[3] As a theoretical basis for future experiments of the qBounce Collaboration[3,4] on an experimental analysis of effects of Lorentz-invariance violation in the neutron sector of the SME, we analyze corrections to the transition frequencies between quantum gravitational states of unpolarized and polarized UCNs and estimate upper bounds on parameters of Lorentz-invariance violation using a current sensitivity of $\Delta E < 2 \times 10^{-15}$ eV.[4] As an example, we give a numerical analysis of the corrections to the transition frequencies of transitions between the ground $|1\rangle$ and the third excited $|4\rangle$ quantum gravitational state of unpolarized and polarized UCNs.

2. Nonrelativistic hamiltonian for Lorentz violation

The nonrelativistic hamiltonian correction Φ_{nLV} due to the violation of Lorentz invariance has been derived from the relativistic SME lagrangian by Kostelecký and Lane[5] using the Foldy–Wouthuysen canonical transformation:[1,5]

$$
\begin{aligned}
\Phi_{\mathrm{nLV}} = {}& -2\tilde{b}_j^n S_j - \left(2c_{jk}^n + c_{00}^n \delta_{jk}\right)\frac{p_j p_k}{2m} + \Big[\left(4d_{0j}^n + 2d_{j0}^n - \varepsilon_{jmn}g_{mn0}^n\right)\delta_{k\ell} \\
& + \varepsilon_{\ell mn}g_{mn0}^n \delta_{jk} - 2\varepsilon_{j\ell m}\left(g_{m0k}^n + g_{mk0}^n\right)\Big]\frac{p_j p_k}{2m} S_\ell + \dots,
\end{aligned}
\tag{1}
$$

where we have used the notation $\tilde{b}_j^n = b_j^n - md_{j0}^n + \frac{1}{2}m\varepsilon_{jk\ell}g_{k\ell0}^n - \frac{1}{2}\varepsilon_{jk\ell}H_{k\ell}^n$.[6] Here, $p_j = -i\nabla_j$ and $S_j = \frac{1}{2}\sigma_j$ are the neutron three-momentum and spin operators, respectively, and σ_j are the 2×2 Pauli matrices. The ellipsis denotes terms whose contribution to the transition frequencies can be neglected.[4]

The evolution of quantum gravitational states of UCNs is described by the Schrödinger–Pauli equation

$$
i\frac{\partial \Psi_{k\sigma}}{\partial t} = \mathrm{H}\Psi_{k\sigma}, \quad \mathrm{H} = -\frac{1}{2m}\Delta + mgz + \Phi_{\mathrm{nLV}},
\tag{2}
$$

where $\Psi_{k\sigma}$ is the two-component spinorial wave function of UCNs in the kth gravitational state and in the spin eigenstates $\sigma = \uparrow$ or $\downarrow$, Δ is the Laplace operator, and mgz is the gravitational potential of the Earth.

3. Standard laboratory and canonical Sun-centered frames

The experiments with UCNs bouncing in the gravitational field of the Earth are being performed in a laboratory at the Institut Laue–Langevin (ILL) in Grenoble. The ILL laboratory is fixed to the surface of the Earth in the northern hemisphere (see Fig. 1). The beam of UCNs moves from south to north antiparallel to the x direction and with UCN energies quantized in the z direction.

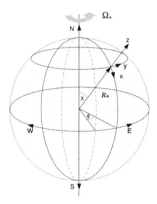

Fig. 1: Position of the ILL laboratory of the qBounce experiments on the surface of the Earth.

The transition frequency for the transition $|q\rangle \rightarrow |p\rangle$ is defined as $\nu_{pq} = (E_p - E_q)/2\pi$, where E_p and E_q are binding energies of the quantum gravitational states $|p\rangle$ and $|q\rangle$, respectively. The energy levels of quantum gravitational states of unpolarized UCNs are two-fold degenerate, and for the calculation of corrections one has to use the stationary perturbation theory of degenerate bound states.[7] However, as has been shown in Ref. 4, the splittings of the energy levels can be neglected in comparison with the shifts. In turn, for the calculation of the corrections to the transition frequencies of non-spin-flip and spin-flip transitions of polarized UCNs, when UCNs are in definite spin eigenstates, one may use stationary perturbation theory for bound states without degeneracy.[4] In a laboratory on the surface of the Earth, such as the ILL, parameters for Lorentz-invariance violation should vary in time with a period related to the Earth's rotation. In order to avoid such a time dependence, parameters for Lorentz-invariance violation should be defined in an inertial frame, for example, in the canonical Sun-centered frame, which remains approximately inertial over thousands of years.[6] The upper bounds of parameters of Lorentz-invariance violation, defined in the canonical Sun-centered frame and obtained in Ref. 4, are adduced in Table 1, where we have used the notation $\tilde{d}_Z^n = m(\bar{d}_{TZ}^n + \frac{1}{2}\bar{d}_{ZT}^n) - \frac{1}{2}\bar{H}_{XY}^n$.[6,8] They are represented in the form accepted in Ref. 8. Parameters violating Lorentz invariance are defined by vacuum terms and fluctuations.[1] In Table 1 we have taken into account the contributions of vacuum terms only, whereas contributions of fluctuations have been neglected.[9,10] The results, represented in Table 1 and obtained

Table 1: Upper bounds on some parameters for Lorentz-invariance violation in the neutron sector of the SME estimated at the sensitivity $\Delta E < 2 \times 10^{-15}$ eV.[4]

Combination	Result
$|\bar{c}_{XX}^n|$	$< 4.4 \times 10^{-4}$
$|\bar{c}_{YY}^n|$	$< 4.4 \times 10^{-4}$
$|\bar{c}_{ZZ}^n|$	$< 4.4 \times 10^{-4}$
$|\bar{c}_{TT}^n|$	$< 1.3 \times 10^{-3}$
$|\bar{g}_{XTY}^n - \bar{g}_{YTX}^n|$	$< 10^{-4}$
$|\tilde{g}_Q|$	$< 1.3 \times 10^{-4}$ GeV
$|\tilde{d}_Z^n + \frac{1}{2} H_{XY}|$	$< 2.3 \times 10^{-5}$ GeV
$|\tilde{b}_Z^n|$	$< 1.4 \times 10^{-24}$ GeV

for the current sensitivity $\Delta E < 2 \times 10^{-15}$ eV of the qBounce experiments,[3] can be treated as a theoretical basis for experimental searches of effects of Lorentz-invariance violation in the experiments of the qBounce Collaboration.[3,4]

Acknowledgments

This work was supported by the Austrian "Fonds zur Förderung der Wissenschaftlichen Forschung" (FWF) under contracts P31702-N27, P26636-N20, and the German "Deutsche Forschungsgemeinschaft" (DFG) under contract AB128/5-2, and by the MA 23 (FH-Call 16) under project "Photonik - Stiftungsprofessur für Lehre."

References

1. D. Colladay and V.A. Kostelecký, Phys. Rev. D **58**, 116002 (1998).
2. H. Abele, T. Jenke, D. Stadler, and P. Geltenbort, Nucl. Phys. A **827**, 593c (2009).
3. G. Cronenberg et al., Nat. Phys. **14**, 1022 (2018) and references therein.
4. A.N. Ivanov, M. Wellenzohn, and H. Abele, to appear.
5. V.A. Kostelecký and C.D. Lane, J. Math. Phys. **40**, 6245 (1999).
6. V.A. Kostelecký and C.D. Lane, Phys. Rev. D **60**, 116010 (1999).
7. A.S. Davydov, *Quantum Mechanics*, Pergamon Press, Oxford, 1965, p. 178.
8. *Data Tables for Lorentz and CPT Violation*, V.A. Kostelecký and N. Russell, 2019 edition, arXiv: 0801.0287v12.
9. Q.G. Bailey and V.A. Kostelecký, Phys. Rev. D **74**, 045001 (2006).
10. V.A. Kostelecký and J.D. Tasson, Phys. Rev. D **83**, 016013 (2011).

The SME with Gravity and Explicit Diffeomorphism Breaking

R. Bluhm

Department of Physics and Astronomy, Colby College, Waterville, ME 04901, USA

This overview looks at what happens when the Standard-Model Extension is used to investigate gravity theories with explicit diffeomorphism breaking. It is shown that when matter–gravity couplings are included, the Standard-Model Extension generally maintains consistency with the Bianchi identities, and it therefore provides a useful phenomenological framework for investigating the effects of explicit diffeomorphism breaking in gravity theories.

1. Introduction

In the gravity sector of the Standard-Model Extension (SME), it is traditionally assumed that the SME coefficients arise as vacuum expectation values in a process of spontaneous diffeomorphism and local Lorentz breaking.[1] With spontaneous breaking, since all the fields are dynamical, consistency with geometric identities such as the Bianchi identities is assured. Furthermore, it was spontaneous Lorentz breaking that provided one of the original motivations for studying Lorentz breaking, since it was shown that mechanisms in string field theory can lead to this form of spacetime-symmetry breaking.[2]

In addition, when the symmetry breaking is spontaneous, the excitations of the background fields take the form of Nambu–Goldstone (NG) and massive Higgs-like excitations. For example, the diffeomorphism NG excitations take the form of Lie derivatives acting on the vacuum values. With this known form for the NG modes, systematic procedures for developing the post-Newtonian limit of the SME, including matter–gravity couplings, have been carried out and have been used to place experimental bounds on gravitational interactions that break spacetime symmetry.[3,4]

In flat spacetime, when the SME coefficients are assumed to be constant at leading order, energy–momentum conservation holds, and it is not crucial whether the Lorentz violation is spontaneous versus explicit. Instead, the SME coefficients are viewed simply as fixed backgrounds, which can be bounded experimentally. However, when gravity is included in the SME, and the SME coefficients are treated as nondynamical backgrounds that explicitly break diffeomorphisms, then conflicts with the Bianchi identities can arise. It is for this reason, that the spacetime-symmetry breaking is traditionally assumed to be spontaneous when gravity is included.

Nonetheless, gravity theories with explicit spacetime breaking are of interest as modifications of General Relativity. Examples include massive gravity and Hořava gravity. The question of whether the SME can be used to test for signals of spacetime-symmetry breaking in gravity theories with explicit breaking is therefore relevant from both a theoretical and experimental perspective.

2. SME with explicit breaking

To consider the SME with gravity and explicit spacetime symmetry breaking, consider an action of the form

$$S_{\text{SME}} = \int d^4x \sqrt{-g} \left[\frac{1}{2}R + \mathcal{L}_{\text{LI}}(g_{\mu\nu}, f_\sigma) + \mathcal{L}_{\text{LV}}(g_{\mu\nu}, f_\sigma, \bar{k}_{\mu\nu\dots}) \right]. \quad (1)$$

The lagrangian is separated into an Einstein–Hilbert term (with $8\pi G = 1$), a Lorentz-invariant (LI) term, and a Lorentz-violating (LV) term. The field f_σ represents a generic dynamical matter field, while $\bar{k}_{\mu\nu\dots}$ generically denotes an SME coefficient in the context of a theory with explicit breaking. The SME coefficient in this context is a fixed nondyamical background, which has no equations of motion. Its inclusion in the action explicitly breaks diffeomorphism invariance.

If a vierbein formalism is used, components of the background field with respect to a local Lorentz basis can appear in the lagrangian as well. In this case, the background field explicitly breaks local Lorentz invariance in addition to diffeomorphisms, and the consequences of both explicit diffeomorphism breaking and local Lorentz breaking can be explored in the context of Riemann–Cartan geometry.[5] However, for the sake of brevity in this overview, only the SME in Riemann spacetime is considered, and the focus is restricted to explicit breaking of diffeomorphisms.

Despite the fact that S_{SME} is not invariant under diffeomorphisms, the action must still be invariant under general coordinate transformations in order to maintain observer independence. By considering infinitesimal general coordinate transformations, with $x^\mu \to x'^\mu(x) = x^\mu - \xi^\mu$, four identities follow as a result of S_{SME} remaining unchanged.[6] When the matter fields are put on shell, and the Bianchi identities are applied, these identities have

124

the form

$$D_\mu T^{\mu\nu} - \frac{\delta \mathcal{L}_{\rm LV}}{\delta \bar{k}_{\alpha\beta\cdots}} D^\nu \bar{k}_{\alpha\beta\cdots} + g^{\nu\sigma} D_\alpha \left(\frac{\delta \mathcal{L}_{\rm LV}}{\delta \bar{k}_{\alpha\beta\cdots}} \bar{k}_{\sigma\beta\cdots} \right) + \cdots = 0, \qquad (2)$$

where $T^{\mu\nu}$ is the energy–momentum tensor and $\frac{\delta \mathcal{L}_{\rm LV}}{\delta \bar{k}_{\alpha\beta\cdots}}$ represents the Euler–Lagrange expression for the background. In a theory where the background is dynamical, such as when the symmetry breaking is spontaneous, the Euler–Lagrange expression $\frac{\delta \mathcal{L}_{\rm LV}}{\delta \bar{k}_{\alpha\beta\cdots}}$ vanishes on shell, and consistency with covariant energy–momentum conservation follows. However, with explicit breaking, the background is nondynamical and its Euler–Lagrange equations need not hold, which potentially leads to a conflict with the requirement of covariant energy–momentum conservation.

Evidently, consistency with $D_\mu T^{\mu\nu} = 0$ requires the four equations,

$$-\frac{\delta \mathcal{L}_{\rm LV}}{\delta \bar{k}_{\alpha\beta\cdots}} D^\nu \bar{k}_{\alpha\beta\cdots} + g^{\nu\sigma} D_\alpha \left(\frac{\delta \mathcal{L}_{\rm LV}}{\delta \bar{k}_{\alpha\beta\cdots}} \bar{k}_{\sigma\beta\cdots} \right) + \cdots = 0, \qquad (3)$$

to hold on shell even though $\frac{\delta \mathcal{L}_{\rm LV}}{\delta \bar{k}_{\alpha\beta\cdots}} \neq 0$. Interestingly, these equations can in general hold in a theory with explicit breaking, despite the fact that the Euler–Lagrange equations for $\bar{k}_{\mu\nu\cdots}$ do not hold. This is because there are four extra degrees of freedom in the theory due to the loss of local diffeomorphism invariance. As long as these four additional degrees of freedom are not suppressed or decouple, then they can take values that can satisfy (3) on shell.

A Stückelberg approach can be used to reveal the four extra degrees of freedom that occur with explicit breaking. In this approach, four dynamical scalars ϕ^A with $A = 0, 1, 2, 3$, are introduced through the substitution $\bar{k}_{\mu\nu\cdots}(x) \to D_\mu \phi^A D_\nu \phi^B \cdots \bar{k}_{AB\cdots}(\phi)$. Since the scalars are dynamical, diffeomorphism invariance is restored. However, the number of degrees of freedom does not change, since the four local symmetries can be used to eliminate the four scalars. In particular, fixing the scalars as $\phi^A = \delta^A_\mu x^\mu$ gives back the original explicit-breaking theory in terms of $\bar{k}_{\mu\nu\cdots}(x)$.

By expanding about the fixed values of the scalars, letting $\phi^A = \delta^A_\mu(x^\mu + \pi^\mu)$, the role of the four additional excitations, denoted as π^μ, can be examined. What is found is that the resulting Stückelberg excitations take the same form as in a Lie derivative acting on the background,

$$\mathcal{L}_\xi \bar{k}_{\mu\nu} \simeq (D_\mu \pi^\alpha) \bar{k}_{\alpha\nu\cdots} + (D_\nu \pi^\alpha) \bar{k}_{\mu\alpha\cdots} + \cdots + \pi^\alpha D_\alpha \bar{k}_{\mu\nu\cdots}. \qquad (4)$$

What this means is that the four extra degrees of freedom occurring with explicit breaking have the same form as the NG modes in a corresponding theory with spontaneous breaking. Effectively, the Stückelberg approach introduces only the four NG modes that are minimally required to maintain consistency, while no massive Higgs-like excitations are generated.

In the pure-gravity sector of the SME in a linearized post-Newtonian limit, the explicit-breaking background $\bar{k}_{\mu\nu...}$ couples to the linearized curvature tensor $R^{\kappa}{}_{\lambda\mu\nu}$. If the Stückelberg approach is used in this case, then the NG modes, π^{μ}, couple to the linearized curvature tensor and its contractions. However, the linearized curvature tensor is invariant under diffeomorphisms at leading order. This means the NG modes can be gauged away while $R^{\kappa}{}_{\lambda\mu\nu}$ does not change. Hence, in the pure-gravity sector, a useful post-Newtonian limit does not result, since the NG modes decouple at leading order.[5]

However, when matter–gravity couplings are included, the background SME tensors $\bar{k}_{\mu\nu...}$ have additional couplings to matter terms in the action that are not invariant under diffeomorphisms at leading order. Thus, in this case, the NG modes do not decouple, and the same procedures that were carried out in the case of spontaneous breaking[4] can therefore be used as well when the symmetry breaking is explicit.

3. Conclusion

In the SME with gravity, when diffeomorphisms are explicitly broken by couplings to a nondynamical background SME coefficient $\bar{k}_{\mu\nu...}$, the Stückelberg approach reveals that the four extra degrees of freedom that result from the symmetry breaking have the same form as the four NG modes that appear in a corresponding theory with spontaneous breaking. As long as these modes are not suppressed or do not decouple, as is generally the case in a linearized treatment that includes matter–gravity couplings, then the same procedures can be used with explicit breaking that were developed assuming spontaneous breaking. In particular, the SME can be used to search for potential experimental signals that might occur in matter–gravity couplings in gravity theories with explicit spacetime-symmetry breaking.

References

1. V.A. Kostelecký, Phys. Rev. D **69**, 105009 (2004).
2. V.A. Kostelecký and S. Samuel, Phys. Rev. D **39**, 683 (1989).
3. Q.G. Bailey and V.A. Kostelecký, Phys. Rev. D **74**, 045001 (2006).
4. V.A. Kostelecký and J.D. Tasson, Phys. Rev. D **83**, 016013 (2011).
5. R. Bluhm, Phys. Rev. D **91**, 065034 (2015).
6. R. Bluhm and A. Šehić, Phys. Rev. D **94**, 104034 (2016).

Standard-Model Extension Constraints on Lorentz and CPT Violation from Optical Polarimetry of Active Galactic Nuclei

Andrew S. Friedman, David Leon, Roman Gerasimov, Kevin D. Crowley,
Isaac Broudy, Yash Melkani, Walker Stevens, Delwin Johnson, Grant Teply,
David Tytler, Brian G. Keating

*Center for Astrophysics and Space Sciences, University of California,
San Diego, La Jolla, CA 92093, USA*

Gary M. Cole

Western Nevada College, Carson City, NV 89703, USA

Vacuum birefringence from Lorentz and CPT violation in the Standard-Model Extension can be constrained using ground-based optical polarimetry of extra-galactic sources. We describe results from a pilot program with an automated system that can perform simultaneous optical polarimetry in multiple pass-bands on different telescopes with an effective 0.45 m aperture.[1] Despite the limited collecting area, our polarization measurements of AGN using a wider effective optical passband than previous studies yielded individual line-of-sight constraints for Standard-Model Extension mass dimension $d = 5$ operators within a factor of about one to ten of comparable broadband polarimetric bounds obtained using data from a 3.6 m telescope with roughly 64 times the collecting area.[2] Constraining more general anisotropic Standard-Model Extension coefficients at higher d would require more AGN along different lines of sight. This motivates a future dedicated ground-based, multi-band, optical polarimetry AGN survey with $\gtrsim 1$ m-class telescopes, to obtain state-of-the-art anisotropic Standard-Model Extension $d = 4, 5, 6$ constraints, while also using complementary archival polarimetry. This could happen more quickly and cost-effectively than via spectropolarimetry and long before more competitive constraints from space- or balloon-based x-ray/γ-ray polarization measurements.

The Standard-Model Extension (SME)[3] allows for the violation of both Lorentz and CPT symmetry. Some SME coefficients predict vacuum bire-fringence, resulting in a wavelength-dependent rotation of the plane of linear polarization for photons. Such SME effects would increasingly depolarize light traveling over cosmological distances, with stronger observable effects predicted at higher redshifts and higher energies. Broadband polarimetry and spectropolarimetry of high-redshift extragalactic sources can thus be used to place increasingly sensitive astrophysical bounds on the SME.

Since SME effects can vary across the sky, one requires multiple measurements along different lines of sight to adequately constrain the most natural anisotropic SME models: the number $N(d)$ of distinct anisotropic vacuum-birefringent SME coefficients increases according to $N(d) = (d-1)^2$ and $N(d) = 2(d-1)^2 - 8$ for odd and even mass dimension d, respectively. Thus, the mass dimensions $d = 5, 7, 9, \ldots$ and $d = 4, 6, 8, 10, \ldots$, require respective polarization measurements along at least $N(d) = 16, 36, 64, \ldots$ or $10, 42, 90, 154, \ldots$ independent lines of sight for full coefficient coverage.

Space- or balloon-based x-ray/γ-ray polarimetry can yield very sensitive line-of-sight constraints on the SME.[4] However, there are currently only about ten published x-ray/γ-ray polarization measurements of gamma-ray bursts (GRBs) that are not upper/lower limits.[4,5] By contrast, there are thousands of AGN with published broadband optical polarimetry and hundreds with spectropolarimetry.[7,8] At present, it is thus much more feasible to obtain wider sky coverage quickly, including many sources over a range of redshifts, by analyzing archival polarimetry of the most highly polarized AGN, including BL Lacs, Blazars, and highly polarized quasars.

Ultimately, advances in space- or balloon-based x-ray/γ-ray polarimetry of high-redshift transient GRBs could provide significantly stronger bounds on anisotropic SME coefficients than optical AGN polarimetry in the coming decades.[2,4,6,9] However, since ground-based optical polarimetry and spectropolarimetry have smaller statistical and systematic errors than the more expensive and difficult x-ray/γ-ray polarimetry measurements, optical studies of AGN—the brightest, continuous, highly polarized, extragalactic optical sources—represent the most cost-effective approach to improve constraints on anisotropic SME coefficients today.[1,2,10]

One can test CPT-odd birefringent SME coefficients with broadband polarimetry as follows.[1-3] The $k_{(V)jm}^{(d)}$ SME coefficients predict a rotation of the linear polarization plane. For two photons with energies $E_1 < E_2$ emitted in the rest frame of a source at redshift z with the same polarization, the difference in polarization angle observed on Earth is

$$\Delta \psi^{(d)}(z) \approx \left(E_2^{d-3} - E_1^{d-3} \right) L^{(d)}(z) \sum_{jm} Y_{jm}(\theta, \phi) k_{(V)jm}^{(d)}, \qquad (1)$$

where $Y_{jm}(\theta, \phi)$ are the spin-weighted spherical harmonics with celestial coordinates (θ, ϕ), and $L^{(d)}(z) = \int_0^z \frac{(1+z')^{d-4}}{H(z')} dz'$ is the effective comoving distance; $H(z)$ is the Hubble parameter for a FRW cosmology.

If we conservatively assume 100% intrinsic source polarization at all wavelengths,[1,2,10] integrating Eq. (1) over the effective energy bandpass

$T(E)$ yields intensity-normalized Stokes parameters $q^{(d)}(z)$ and $u^{(d)}(z)$ via

$$q^{(d)}(z) + iu^{(d)}(z) = \int_{E_1}^{E_2} \exp\left[i\left(E^{d-3} - E_1^{d-3}\right)\xi(z)\right] T(E)\, dE, \quad (2)$$

where $\xi(z) \equiv 2L^{(d)}(z)\bar{k}_{(V)}^{(d)}$ and $\bar{k}_{(V)}^{(d)} \equiv \sum_{jm} Y_{jm}(\theta, \phi)k_{(V)jm}^{(d)}$.

The polarization $p_{\max}^{(d)}(z) = \left([q^{(d)}(z)]^2 + [u^{(d)}(z)]^2\right)^{1/2}$ represents the theoretical maximum observable in the SME. Measuring a polarization fraction $p_\star \pm n\sigma_{p_\star} < p_{\max}^{(d)}(z)$ can thus directly yield an n–σ upper bound on the coefficient combination $\bar{k}_{(V)}^{(d)}$. A spherical-harmonic decomposition on the sky can then be used to combine sufficient numbers of line-of-sight constraints to bound all $N(d)$ parameters at a given d.[2,10] These can also include line-of-sight constraints from spectropolarimetry, which can be about two or three orders of magnitude more sensitive than broadband polarimetry for $d = 5$.[2] Tests of CPT-even $d = 4$ SME coefficients are discussed in Ref. 10.

The Array Photo Polarimeter (APPOL) is a pilot program with an automated small telescope system.[1] It can conduct high-cadence, faint-object, optical polarimetry in multiple passbands with polarization-fraction statistical errors $\sigma_p \lesssim 0.5$–1% for targets with visual band magnitude $V \lesssim$ 14–15 mag, and systematic errors $\sigma_p \sim 0.04\%$. These are competitive with the best ground-based optical-telescope measurements. APPOL is located at StarPhysics Observatory in Reno, Nevada at an elevation of 1585 m and serves as a test bed for future polarimeters that could be installed on $\gtrsim 1$ m-class telescopes capable of observing much fainter AGN.[1]

APPOL uses dual-beam inversion optical polarimetry with Savart plate analyzers rotated through a half-wave-plate image sequence;[11] it employs an automated telescope, filter, and instrument-control system with five co-located telescopes on two mounts. We combined simultaneous polarimetry from two co-located Celestron 11- and 14-inch telescopes with an effective 17.8 inch (0.45 m) telescope diameter with Lum and I_c filters into an effective optical bandpass with high transmission over the $\lambda \simeq 400$–900 nm range of the two filters.[1] This yields more stringent SME bounds than either filter alone and achieves the effective collecting power of a larger telescope.

Our initial APPOL campaign observed two sources: BL Lacertae and S5 B0716+714 at redshifts $z = 0.069$ and $z = 0.31$, respectively. This can only give SME line-of-sight constraints or bound the $k_{(V)00}^{(d)}$ isotropic CPT-odd SME coefficient.[1] Simultaneous optical polarimetry with our $Lum + I_c$ filter can yield SME line-of-sight constraints that are theoretically up to ten times ($d = 5$) and 30 times ($d = 6$) more sensitive than in the I_c band alone. Despite our small effective 0.45 m aperture, we achieved $d = 5$ line-of-sight constraints within a factor of up to ten in sensitivity compared to relevant constraints using broadband optical polarimetry with a V-band filter on a 3.6 m telescope with roughly 64 times the collecting area.[1,2]

Using archival optical polarimetry and spectropolarimetry for AGN (and GRB afterglows), there is a unique opportunity to test $k_{(V)}^{(d)}$ SME coefficients about one to two orders of magnitude better than previous work.[2,10] While most archival optical data used a single filter, we conjecture that simultaneous observations in as few as two filters could constrain $k_{(V)}^{(d)}$ more cost-effectively than spectropolarimetry on $\gtrsim 2$ m-class telescopes. We are testing this now by performing simultaneous two-band optical polarimetry on roughly 10–20 of the brightest, highly polarized, AGN that an upgraded 0.5 m APPOL system can reasonably observe. This will enable design-feasibility studies for a ground-based, multi-band, optical polarimetry survey of high-redshift AGN with $\gtrsim 1$ m-class telescopes. Finally, with thousands of sources over the sky, archival and new optical polarimetry could, for the first time, provide sufficient data to constrain not just individual $k_{(V)}^{(d)}$, but also a possible redshift dependence of any SME coefficients, potentially revealing time variation of the underlying fields and elucidating the role of the associated new physics over cosmic history.

Acknowledgments

ASF and BGK acknowledge support from US National Science Foundation INSPIRE Award PHYS 1541160 and UCSD's Ax Center for Experimental Cosmology. DT is supported in part by NSF award AST1413568.

References

1. A.S. Friedman *et al.*, Phys. Rev. D **99**, 035035 (2019).
2. F. Kislat and H. Krawczynski, Phys. Rev. D **95**, 083013 (2017).
3. V.A. Kostelecký and M. Mewes, Phys. Rev. D **80**, 015020 (2009).
4. V.A. Kostelecký and M. Mewes, Phys. Rev. Lett. **110**, 201601 (2013).
5. J. Wei, Mon. Not. R. Astron. Soc. **485**, 2401 (2019).
6. M. Pearce *et al.*, Astropart. Phys. **104**, 54 (2019).
7. D. Sluse *et al.*, Astron. Astrophys. **433**, 757 (2005).
8. P.S. Smith *et al.*, in N. Johnson and D. Thompson, eds., *Proceedings of the 2009 Fermi Symposium*, eConf C0911022, 2009.
9. F. Kislat, talk **109.79** presented at *17th HEAD Meeting of the American Astronomical Society*, Monterey, 17–21 March, 2019.
10. F. Kislat, Symmetry **10**, 596 (2018).
11. J. Tinbergen, *Astronomical Polarimetry*, Cambridge University Press, Cambridge, 2005.

Lorentz and CPT Tests
with Exotic Atoms, Clocks, and Other Systems

Arnaldo J. Vargas

Physics Department, Loyola University New Orleans, New Orleans, LA 70118, USA

Signals for Lorentz violation in atomic spectroscopy experiments, clock-comparison experiments, and spin-precession experiments are presented. The differences in the signals associated with minimal and nonminimal Lorentz-violating fermion terms are discussed.

1. Motivation and introduction

The Standard-Model Extension (SME) is an effective field theory that facilitates systematic tests of CPT and Lorentz symmetry.[1] Some of the first minimal-SME models focused on atomic systems and on spin-precession effects.[2] Experimental analyses of these systems were conducted including atomic-spectroscopy experiments with ordinary and exotic atoms,[3] spin-precession experiments with first- and second-generation particles,[4] and clock-comparison experiments.[5]

Theoretical advances permitted a systematic classification of fermion Lorentz-violating operators of arbitrary mass dimension.[6] This work encouraged the construction of nonminimal Lorentz-violation models for the previously mentioned systems. Nonminimal models for Lorentz violation exist for muon $g-2$ experiments;[7] for Penning-trap experiments with electrons, positrons, protons, and antiprotons;[8] for atomic-spectroscopy experiments with light atoms such as hydrogen,[9] antihydrogen,[9] deuterium,[9] positronium,[9] and muonic atoms;[7] and for clock-comparison experiments with comagnetometers, atomic fountain clocks, atomic masers, ion-trap clocks, and optical-lattice clocks.[10]

In these proceedings, the main signals predicted by atomic-spectroscopy experiments, clock-comparison experiments, and muonic g-2 experiments are presented.

2. Signals for Lorentz violation

A signal for Lorentz violation is a breaking of rotational symmetry. This symmetry can be tested by searching for changes in the atomic spectrum due to the rotation of the atom. The atom can be rotated by using a magnetic field to manipulate the orientation of the atomic magnetic dipole moment. In this section, we will consider only the effects produced by fixing the orientation of an atom or particle with respect to a laboratory frame on the surface of the Earth. As the Earth rotates the system will rotate with it and this rotation can produce sidereal and annual variations in its resonance frequencies in the presence of Lorentz violation.

In the context of the minimal SME, only the first and the second harmonic of the sidereal frequency can contribute to the sidereal variation of any atomic transition frequency. If nonminimal terms are allowed in the model then the harmonics that contribute to the frequency variation are determined by the angular-momentum quantum numbers of the states involved in the transition.[9] The general rule is that the number of harmonics of the sidereal frequency and the number of coefficients for Lorentz violation that contribute to the variation of the frequency increase with the angular momentum.[9] The differences between the minimal and nonminimal Lorentz-violating models are more significant for states with high angular momentum. In the case of the hydrogen atom, the possible signals for Lorentz violation in transitions between low-angular-momentum states, such as the $1s$-$2s$ transition and the Zeeman-hyperfine transitions of the ground state, are the same in both models. However, harmonics higher than the second harmonic of the sidereal frequency can contribute to the variations of the $2s$–np and the $2s$–nd transition frequencies.[9]

Another example is the difference between minimal and nonminimal Lorent-violating models for a ^{129}Xe–^{3}He comagnetometer and a cesium atomic fountain clock. The signals for Lorentz violation in the comagnetometer are the same in both models as it uses transitions involving low-angular-momentum states.[10] The atomic fountain clock uses states with higher atomic angular momentum than the comagnetometer. For that reason, the nonminimal model for the cesium clock allows contributions from higher harmonics and more coefficients for Lorentz violation than the minimal model.[10]

These differences between minimal and nonminimal models are also relevant in antihydrogen spectroscopy experiments. Several collaborations at CERN have measured[11] or are planning[12,13] to measure the $1s$–$2s$ and the ground-state hyperfine transitions in antihydrogen. These transitions were targeted because of the prospects of high-precision spectroscopy. Indeed, models for CPT and Lorenz violation show that these are the most sensitive antihydrogen transitions to certain CPT-violating operators.[9] The same models also show that any systematic study of CPT violation with antihydrogen spectroscopy should not be limited to transitions between s states as these transitions are sensitive only to a small fraction of the CPT-violating operators that could be studied with antihydrogen spectroscopy.[9]

Lorentz-violating effects can produce energy shifts similar to the Zeeman and Stark effects. Transitions that are insensitive to these effects might also be insensitive to leading-order Lorentz violation. This is the case for the clock transitions in hydrogen masers[9] and cesium atomic-fountain clocks.[10] One implication of this result is that hydrogen masers and fountain clocks can be used as frequency references in tests of Lorentz and CPT symmetry as they are insensitive to leading-order Lorentz violation. For frequencies sensitive to the applied fields, different methods are used to reduce the uncertainties generated by the fields, and depending on the method Lorentz violation might be observable or not as discussed in Ref. 10.

Any systematic search for signals of Lorentz violation needs to consider the possibility that different particle species could couple differently with the Lorentz-violating background fields. This is the main motivation to study Lorentz violation with muon $g - 2$ experiments and with muonic atoms such as muonium and muonic hydrogen.[7] Even in ordinary-matter experiments, we need to consider the sensitivities of the experiments to the electron, proton, and neutron Lorentz-violating operators. For example, low-frequency transitions, such as hyperfine Zeeman transitions, usually produce tighter constraints on Lorentz-violating effects than other atomic transitions.[9,10] However, these transitions tend to be sensitive only to the nucleon Lorentz-violating operators and not to the electron Lorentz-violating operators. Optical transitions are the best candidates to study electron Lorentz-violating operators, despite being less sensitive to Lorentz violation than hyperfine Zeeman transitions because they receive contributions from the electron Lorentz-violating operators.[10]

The minimal and nonminimal coeffcients depend differently on the momentum and energy of the particles. For example, performing the same experiment with muonic hydrogen instead of muonium will enhance the sensitivity to some nonminimal terms because the muon has a greater linear momentum in muonic hydrogen than the antimuon in muonium.[7] Another example is that the new measurement of the antimuon anomalous frequency at Fermilab will use more energetic antimuons compared to the experiment at J-PARC[14] and for that reason it will be more sensitive to nonminimal Lorentz-violating operators than the $g - 2$ experiment at J-PARC.[7]

Acknowledgments

This work was supported by the Department of Energy under grant No. DE-SC0010120 and by the Indiana University Center for Spacetime Symmetries.

References

1. D. Colladay and V.A. Kostelecký, Phys. Rev. D **55**, 6760 (1997); Phys. Rev. D **58**, 116002 (1998); V.A. Kostelecký, Phys. Rev. D **69**, 105009 (2004).
2. R. Bluhm, V.A. Kosteleck, and N. Russell Phys. Rev. D **57**, 3932 (1998); R. Bluhm, V.A. Kostelecký, and N. Russell, Phys. Rev. Lett. **82**, 2254 (1999); V.A. Kostelecký and C.D. Lane, Phys. Rev. D **60**, 116010 (1999); R. Bluhm, V.A. Kostelecký, and C.D. Lane, Phys. Rev. Lett. **84**, 1098 (2000); R. Bluhm, V.A. Kostelecký, C.D. Lane, and N. Russell, Phys. Rev. Lett. **88**, 090801 (2002).
3. D.F. Phillips *et al.*, Phys. Rev. D **63**, 111101(R) (2001); V.W. Hughes *et al.*, Phys. Rev. Lett. **87**, 111804 (2001); A. Matveev *et al.*, Phys. Rev. Lett. **110**, 230801 (2013).
4. G.W. Bennett *et al.*, Phys. Rev. Lett. **100**, 091602 (2008); A. Mooser *et al.*, Nature **509**, 596 (2014); H. Nagahama *et al.*, Nat. Commun. **8**, 14084 (2017).
5. F. Canè *et al.*, Phys. Rev. Lett. **93**, 230801 (2004); M. Smiciklas, J.M. Brown, L.W. Cheuk, and M.V. Romalis, Phys. Rev. Lett. **107**, 171604 (2011); F. Allmendinger *et al.*, Phys. Rev. Lett. **112**, 110801 (2014); H. Pihan-Le Bars *et al.*, Phys. Rev. Lett. **95**, 075026 (2017); C. Sanner *et al.*, Nature **567**, 204 (2019).
6. V.A. Kostelecký and M. Mewes, Phys. Rev. D **88**, 096006 (2013); M. Mewes, these proceedings.
7. A.H. Gomes, V.A. Kostelecký, and A.J. Vargas, Phys. Rev. D **90**, 076009 (2014).
8. Y. Ding and V.A. Kostelecký, Phys. Rev. D **94**, 056008 (2016); Y. Ding, these proceedings.
9. V.A. Kostelecký and A.J. Vargas, Phys. Rev. D **92**, 056002 (2015).
10. V.A. Kostelecký and A.J. Vargas, Phys. Rev. D **98**, 036003 (2018).
11. M. Ahmadi *et al.*, Nature **548**, 66 (2017); M. Ahmadi *et al.*, Nature **557**, 71 (2018); T. Friesen, these proceedings.
12. E. Widmann *et al.*, Hyperfine Int. **215**, 1 (2013); M. Simon, these proceedings; A. Nanda, these proceedings.
13. G. Gabrielse *et al.*, Phys. Rev. Lett. **89**, 213401 (2002).
14. R.M. Carey *et al.*, Fermilab proposal 0989, 2009; M. Aoki *et al.*, J-PARC proposal J-PARC-PAC2009-12, 2009.

Lorentz-Violation Constraints with Astroparticle Physics

H. Martínez-Huerta

Instituto de Física de São Carlos, Universidade de São Paulo,
Av. Trabalhador São-carlense 400, CEP 13566-590 São Carlos, SP, Brazil

Astroparticle physics has recently reached a new level of precision due to the construction of new observatories operating innovative technologies and the detection of large numbers of events and sources. Precise measurements of cosmic and gamma rays can be used for testing fundamental physics, such as Lorentz symmetry. Although Lorentz-violation signatures are expected to be small, the high energies and long distances involved in astrophysical observations lead to an unprecedented opportunity for this task. In this summary, exclusion-limit results are presented from different types of astrophysical Lorentz tests through generic modification of the particle dispersion relation in the photon sector through pair-production threshold shifts and photon decay. Some perspectives for the next generation of gamma-ray telescopes are also given.

1. MDR for astroparticle tests

The introduction of a Lorentz-violating term in the lagrangian[1] or spontaneous Lorentz-symmetry breaking[2] can induce modifications to the particle dispersion relation (MDR). A phenomenological generalization of these Lorentz-violating (LV) effects converges on the introduction of a general function of the energy and the momentum. Although there are several forms of MDRs for different particles and underlying LV theories, some of them may lead to similar phenomenology, which can be useful for LV tests in extreme environments, such as astroparticle scenarios. In this context, a family of MDRs can be addressed by the following expression in natural units:

$$E_a^2 - p_a^2 = m_a^2 \pm |\delta_{a,n}| A_a^{n+2}, \tag{1}$$

where a stands for the particle type, A may be either E or p, $\delta_{a,n}$ is the LV parameter, and n is the leading order of the correction from the underlying theory. In some effective field theories, $\delta_{a,n} = \epsilon^{(n)}/M$, where M is the energy scale of the new physics, such as the Plank-energy scale, E_{Pl}, or some quantum-gravity energy scale, E_{QG}, and $\epsilon^{(n)}$ are LV coefficients.

2. Gamma-ray attenuation including LV

Gamma rays from distant sources are significantly attenuated due to the interaction with background light (BL) through pair production $\gamma \gamma_b \longrightarrow e^+ e^-$, where γ_b is a BL photon. This limits how far photons can travel without being absorbed.[3] The physics derived from Eq. (1) can lead to shifts in the minimum BL energy that allows pair production:[4]

$$E_{\gamma_b}^{th} = \frac{m_e^2}{4 E_\gamma K (1-K)} - \tfrac{1}{4} \delta_n^{tot} \, E_\gamma^{n+1},$$ (2)

where $\delta_n^{\text{tot}} = \delta_{\gamma,n} - \delta_{+,n} K^{n+1} - \delta_{-,n} (1-K)^{n+1}$ is a linear combination of LV coefficients from the involved particle species, and K stands for the inelasticity,[4,5] $E_+ = K(E_\gamma + E_{\gamma_b})$. The cumulative effect of this phenomenon results in measurable changes in the expected attenuation of the gamma-ray flux due to BL.[6,7] Due to its nature in the universe, the dominant BL component depends on the gamma-ray energy window. For instance, for $E_\gamma < 10^{14.5}$ eV, the extragalactic BL is dominant, whereas for $10^{14.5}$ eV $< E_\gamma < 10^{19}$ eV it is the CMB. In the extragalactic BL energy region, the subluminal ($\delta_n^{tot} < 0$) LV effect forecast a recovery in the spectrum of TeV-sources that can be measured by current gamma-ray telescopes.[8] A new search for this LV signature with the most updated TeV gamma-ray data set involved 111 measured energy spectra from 38 sources.[6] This data set was shown to be best described by the Lorentz-invariant model, and the stringent limits at 2σ c.l. in Table 1 were established,[6] which are the first published results robust under several tested systematic uncertainties.

The Cherenkov Telescope Array (CTA) is the next-generation ground-based IACT observatory for gamma rays at very high energies with detection capabilities in the energy range from 20 GeV to over 300 TeV with unprecedented precision in energy and directional reconstruction, which will generate an unparalleled opportunity for LV tests.[9] Preliminary results from simulations of CTA observations of the spectrum of the nearby blazar Mrk 501 have shown that CTA will be sensitive to this type of signatures with at least the LV values in Table 1.[10]

A recent search for LV signatures in the CMB-dominant background region considered the so-called GZK-photon flux on Earth.[7] For the first time, this flux was computed employing several ultrahigh-energy cosmic ray (UHECR) injection and source distribution models including the model combination that was shown to best describe the energy spectrum, composition, and arrival direction of UHECR data. This approach corresponds to a source-distribution model that follows a GRB rate evolution, with a power-law energy-spectrum injection model at the source with a rigidity cutoff and including five different species of primary cosmic-ray nuclei.[11] The predicted LV effect was the increase in the predicted GZK-photon flux.[5,7] By comparison of this model prediction with the most updated upper limits to the integrated photon flux obtained by the Pierre Auger Observatory,[12] the LV scenarios in Table 1 were excluded.[7]

3. Photon decay

The phenomenology derived from Eq. (1) can also be used to study super-luminal phenomena predicted in some LV scenarios, such as photon decay, photon splitting, and vacuum Cherenkov radiation.[4,13] The resulting rates for photon decay into an electron–positron pair are very fast and effective once the process is allowed, which suggests an abrupt cutoff in the gamma-ray spectrum with no high-energy photons reaching Earth from cosmological distances above a given threshold. For a given n in Eq. (1), this threshold is[4] $\delta_{\gamma,n} \geq 4m^2/(E_\gamma^n(E_\gamma^2 - 4m^2))$. Using $E_\gamma = 56\,\mathrm{TeV}$ observations at $\sim 5\sigma$ c.l. from the Crab Nebula as reported by the HEGRA telescope, the exclusion limits in Table 1 can be extracted.[4]

Recent results involving gamma-ray energies above 100 TeV by the High Altitude Water Cherenkov (HAWC) observatory,[14] together with the recent development of an energy-reconstruction algorithm, new and stringent constraints on the order of those in Table 1 are being established.[15] In addition, prospects to test photon decay through the observation of the SNR RX J1713.7-394 were reported in the science-motivation paper for the Southern Gamma-Ray Survey Observatory (SGSO), which will be a next-generation wide-field-of-view gamma-ray survey instrument, sensitive to gamma rays

Table 1: LV limits. PP denotes pair production and PD photon decay. The superscript † labels the analysis in the astrophysical scenario that best describes UHECR data. This analysis has also placed a limit on $|\delta_0|$ at the 10^{-20} level. We have taken $\delta_n = (E_{\mathrm{LV}}^{(n)})^{-n} \approx (E_{\mathrm{QG}}^{(n)})^{-n}$ and $E_{\mathrm{Pl}} \approx 1.22 \times 10^{28}$ eV. The HAWC limit is preliminary, and the CTA and SGSO are projected limits based on Refs. 10,16, respectively.

Type	$\|\delta_1\|$ $10^{-30}/\mathrm{eV}$	$\|\delta_2\|$ $10^{-43}/\mathrm{eV}^2$	$E_{\mathrm{LV}}^{(1)}$ $10^{28}\mathrm{eV}$	$E_{\mathrm{LV}}^{(2)}$ $10^{21}\mathrm{eV}$	Threshold Bound	Ref.
Limit	8.3	1.8	12.08	2.38	PP ($\delta_n^{tot} < 0$)	[6]
Limit†	$\sim 10^{-8}$	$\sim 10^{-13}$	$\sim 10^{10}$	$\sim 10^{7}$	PP ($\delta_n^{tot} < 0$)	[7]
CTA	~ 82	~ 11	~ 1.22	~ 0.97	PP ($\delta_n^{tot} < 0$)	[10]
Limit	6.7	1.3	15.0	2.8	PD ($\delta_{\gamma,n} > 0$)	[4]
HAWC prelim.	$\lesssim 0.1$	$\lesssim 0.01$	$\gtrsim 10^{3}$	$\gtrsim 10$	PD ($\delta_{\gamma,n} > 0$)	[15]
SGSO	~ 1	~ 0.1	$\sim 10^{2}$	~ 10	PD ($\delta_{\gamma,n} > 0$)	[16]

in the energy range from $100\,\text{GeV}$ to hundreds of TeV, which can lead to exclusion limits on the order of those in Table 1.[16]

4. Conclusions

The precise measurements of cosmic and gamma rays can be used to test fundamental physics, such as LV. New exclusion limits derived from the effect of shifting the energy threshold of pair production and the instability of photons, compatible with updated data of TeV sources, UHECR, and gamma rays, were presented. Updates and new studies can be expected with the advent of new and better data from these cosmic messengers provided by current experiments, such as HAWC, and next-generation ones, such as CTA and SGSO.

Acknowledgments

The author acknowledges FAPESP support under grant Nos. 2015/15897-1 and 2017/03680-3 as well as the National Laboratory for Scientific Computing (LNCC/MCTI, Brazil) for providing HPC resources of the SDumont supercomputer (https://sdumont.lncc.br).

References

1. S.R. Coleman and S.L. Glashow, Phys. Lett. B **405**, 249 (1997); Phys. Rev. D **59**, 116008 (1999).
2. D. Colladay and V.A. Kostelecký, Phys. Rev. D **58**, 116002 (1998).
3. A. De Angelis *et al.*, Mon. Not. Roy. Astron. Soc. **432**, 3245 (2013).
4. H. Martínez-Huerta and A. Pérez-Lorenzana, Phys. Rev. D **95**, 063001 (2017).
5. H. Martínez-Huerta *et al.*, PoS BHCB **2018**, 010 (2018).
6. R.G. Lang *et al.*, Phys. Rev. D **99**, 043015 (2019).
7. R.G. Lang *et al.*, Astrophys. J. **853**, 23 (2018).
8. J. Biteau and D.A. Williams, Astrophys. J. **812**, 60 (2015); H.E.S.S. Collaboration, H. Abdalla *et al.*, Astrophys. J. **870**, 93 (2019).
9. CTA Consortium, B.S. Acharya *et al.*, *Science with the Cherenkov Telescope Array*, World Scientific, Singapore, 2019.
10. CTA Consortium, F. Gat *et al.*, PoS ICRC **2017**, 623 (2018).
11. M. Unger *et al.*, Phys. Rev. D **92**, 123001 (2015).
12. Pierre Auger Collaboration, C. Bleve, PoS ICRC **2015**, 1103 (2016); Pierre Auger Collaboration, A. Aab *et al.*, JCAP **1704**, 009 (2017).
13. K. Astapov *et al.*, JCAP **1904**, 054 (2019).
14. HAWC Collaboration, A.U. Abeysekara *et al.*, arXiv:1905.12518.
15. HAWC Collaboration, J.T. Linnemann, these proceedings.
16. The SGSO Alliance, A. Albert *et al.*, arXiv:1902.08429.

Renormalization in Nonminimal Lorentz-Violating Field Theory

J.R. Nascimento,[*] A.Yu. Petrov,[*] and C.M. Reyes[†]

[*]Departamento de Física, Universidade Federal da Paraíba,
João Pessoa, Paraíba, Caixa Postal 5008, 58051-970, Brazil

[†]Departamento de Ciencias Básicas, Universidad del Bío-Bío,
Chillán, Casilla 447, Chile

We provide the first step towards renormalization in a nonminimal Lorentz-violating model consisting of normal scalars and modified fermions with mass-dimension five operators. We compute the radiative corrections corresponding to the scalar self-energy, we show that some divergencies are improved, and in the scalar sector they are finite. The pole mass of the scalar two-point function is found and shown to lead to modifications of asymptotic states.

1. Introduction

The Standard-Model Extension (SME) can be regarded as an effective generalized framework to accommodate possible effects of suppressed CPT and Lorentz violation. The SME comprises two different sectors: the minimal sector with operators of mass dimension lower or equal to four,[1] and the nonminimal sector with operators of higher mass dimension.[2] Several bounds have been given using ultrahigh-sensitivity experiments, which have allowed tests of various predictions of Lorentz breakdown in the Standard Model of particle physics and gravity.[3]

A key feature that distinguishes nonminimal models of Lorentz violation from minimal ones is the indefinite metric that arises due to the higher-time-derivative terms. The indefinite metric, as is well known, introduces a pseudo-unitarity relation for the S matrix, which can imply the loss of conservation of probability. Recently, it has been shown that by adopting a Lee–Wick prescription,[4] it is possible to have a consistent unitary theory.[5]

The renormalization of quantum field theories incorporating Lorentz violation can be modified due to radiatively induced operators having different structures not initially present in the original lagrangian.[6] As a general statement, one can say that one-loop Lorentz-violating corrections may have a more dramatic effect on asymptotic states than they do in the typical case. The modified asymptotic space has been mainly shown in computations for the mass pole of two-point functions. In this work, we study the renormalization with particular focus on finite radiative corrections associated to nonminimal terms and the effect of modifying the renormalization conditions due to Lorentz violation.

2. The Yukawa-like model

Consider the Lagrange density[7]

$$\mathcal{L} = \tfrac{1}{2}\partial_\mu\phi\partial^\mu\phi - \tfrac{1}{2}M^2\phi^2 + \bar{\psi}\left(i\slashed{\partial} - m - \bar{\alpha}m\slashed{n}\right)\psi + g_2\bar{\psi}\slashed{n}(n\cdot\partial)^2\psi + g\bar{\psi}\phi\psi, \quad (1)$$

where n^μ is a preferred four-vector, g_2 is a constant with dimension of $(\text{mass})^{-1}$, and g is a typical Yukawa coupling. As a first example of quantum corrections in our model, we study the contribution with two external scalar legs depicted in Fig. 1.

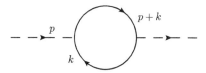

Fig. 1: Scalar self-energy loop diagram.

This diagram is represented by the integral

$$i\Pi(p) = -\tfrac{g^2}{2}\phi(-p)\phi(p)\int \frac{d^4k}{(2\pi)^4}\frac{\text{Tr}((Q_\mu\gamma^\mu+m)(R_\nu\gamma^\nu+m))}{(Q^2-m^2)(R^2-m^2)} \quad (2)$$

where we define

$$Q_\mu(k) = k_\mu - \bar{\alpha}mn_\mu - g_2 n_\mu(n\cdot k)^2, \quad (3)$$
$$R_\mu(p+k) = p_\mu + k_\mu - \bar{\alpha}mn_\mu - g_2 n_\mu(n\cdot(p+k))^2. \quad (4)$$

We can write the correction to the scalar propagator up to second order in p as

$$\tilde{\Pi}(p) = m^2 q_0 + p^2 q_1 + (n\cdot p)^2 q_n, \quad (5)$$

where

$$q_0 = -\frac{i}{48\pi^2 g_2^2 m^2} + \frac{i}{16\pi^2}\left(2(3\gamma_E - 1) - 0.46\right. \tag{6}$$

$$\left.- 3\ln\left(\frac{g_2^2 m^2}{4}\right)\right),$$

$$q_1 = -\frac{i}{12\pi^2}\left(-5 + 6\gamma_E - 0.46 - 3\ln\left(\frac{g_2^2 m^2}{4}\right)\right), \tag{7}$$

$$q_n = \frac{i}{\pi^2}. \tag{8}$$

The renormalized scalar two-point function is given by

$$(\Gamma_{\phi,R}^{(2)})^{-1} = p^2 - M^2 + \Pi_R(p), \tag{9}$$

with

$$\Pi_R(p) = p^2 A_\phi + m^2 B_\phi + (n \cdot p)^2 C_\phi, \tag{10}$$

where A_ϕ, B_ϕ, and C_ϕ are constants that can be deduced from the expressions (6), (7), (8) being

$$iA_\phi = -2g^2 q_1,$$
$$iB_\phi = -2g^2 q_0,$$
$$iC_\phi = -2g^2 q_n. \tag{11}$$

We consider the ansatz $\bar{P}_\phi^2 = p^2 - M_{\text{ph}}^2 + \bar{y}(n \cdot p)^2$ in terms of the two unknown constants M_{ph} and $\bar{y}$. Both constants can be determined using the renormalization condition

$$(\Gamma_{\phi,R}^{(2)})^{-1}|_{\bar{P}_\phi^2=0} = 0. \tag{12}$$

Hence, from (9) replacing the value of p^2 and using the condition (12), we arrive at the equation

$$0 = M_{\text{ph}}^2 - \bar{y}(n \cdot p)^2 - M^2 + A_\phi\left(M_{\text{ph}}^2 - \bar{y}(n \cdot p)^2\right) + B_\phi m^2 + C_\phi(n \cdot p)^2. \tag{13}$$

Due to the independence of each term, we find the two constants

$$M_{\text{ph}}^2 = \frac{(M^2 - B_\phi m^2)}{(1 + A_\phi)}, \qquad \bar{y} = \frac{C_\phi}{1 + A_\phi}, \tag{14}$$

and in consequence also the scalar pole mass $\bar{P}_\phi^2$ which dictates how asymptotic states propagate. Substituting the above expressions in Eq. (9) and using the normalization condition

$$\frac{d(\Gamma_{\phi,R}^{(2)})^{-1}}{d\bar{P}_\phi^2}|_{\bar{P}_\phi^2=0} = Z_\phi^{-1}, \tag{15}$$

we identify the finite wave-function renormalization constant $Z_\phi^{-1} = 1 + A_\phi$.

3. Conclusions

For nonminimal Lorentz-violating models, one should, in general, expect an indefinite metric leading to a nontrivial structure of poles and a pseudo-unitary relation for the S matrix. Also, the nonstandard structure of radiative corrections in general leads to a modified asymptotic space. For the model we have focused on, we have considered the prescription for locating the poles to be dictated by unitarity-conservation requirements. We have found that some radiative corrections are improved, and in fact they are finite in the scalar sector. The pole extraction for the Yukawa-like model has been successfully provided in the scalar sector.

Acknowledgments

The work of AYuP has been supported by CNPq under project No. 303783/2015-0. CMR acknowledges support by FONDECYT under grant No. 1191553.

References

1. D. Colladay and V.A. Kostelecký, Phys. Rev. D **55**, 6760 (1997); Phys. Rev. D **58**, 116002 (1998).
2. V.A. Kostelecký and M. Mewes, Phys. Rev. D**80**, 015020 (2009); Phys. Rev. D**85**, 096005 (2012); Phys. Rev. D**88**, 096006 (2013).
3. *Data Tables for Lorentz and CPT Violation*, V.A. Kostelecký and N. Russell, 2019 edition, arXiv:0801.0287v12.
4. T.D. Lee and G.C. Wick, Nucl. Phys. B **9**, 209 (1969); T.D. Lee and G.C. Wick, Phys. Rev. D **2**, 1033 (1970).
5. C.M. Reyes, Phys. Rev. D **87**, 125028 (2013); M. Maniatis and C.M. Reyes, Phys. Rev. D **89**, 056009 (2014); C.M. Reyes and L.F. Urrutia, Phys. Rev. D **95**, 015024 (2017).
6. R. Potting, Phys. Rev. D **85**, 045033 (2012); M. Cambiaso, R. Lehnert, and R. Potting, Phys. Rev. D **90**, 065003 (2014).
7. J.R. Nascimento, A.Y. Petrov, and C.M. Reyes, Eur. Phys. J. C **78**, 541 (2018).

Establishing a Relativistic Model for Atomic Gravimeters

Ya-jie Wang, Yu-Jie Tan, and Cheng-Gang Shao

MOE Key Laboratory of Fundamental Physical Quantities Measurements,
Hubei Key Laboratory of Gravitation and Quantum Physics,
PGMF and School of Physics, Huazhong University of Science and Technology,
Wuhan 430074, China

This work establishes a high-precision relativistic theoretical model for atomic gravimeters. The starting point is the study of finite speed-of-light effects based on a coordinate transformation, and the analysis is then further extended to include other special- and general-relativistic effects. This model promotes efforts to test General Relativity with atomic interferometry.

1. Introduction

Since high-precision gravimeters have numerous applications such as in the construction of accurate tide models, the study of geodes, and for tests of General Relativity (GR), it is necessary to develop such devices and the corresponding theoretical models. In gravity measurements, we should therefore not only include Newtonian effects, but also some special and general relativistic effects to improve the accuracy of theoretical models.[1] In general, research regarding relativistic effects in atomic gravimeters can be divided into two aspects: finite speed-of-light (FSL) effects,[2-4] and GR effects.[5,6] Since there exists some disagreement[2-6] and incompleteness[5,6] in this context in the literature, we recalculate these effects and derive a more complete and general expression for them.

2. Idea behind interferometric-phase-shift calculation

According to the working principle of atomic gravimeters, a three-Raman-pulse sequence is usually used to interact with the moving atoms. These pulses split, reflect, and recombine the atomic wave packets. As the evolution of atoms is spatially separated, the interferometric signal carries the information of the gravitational field. Thus, the gravitational acceleration can be derived from the measured interferometric phase shift.

A complete atomic-interferometry system consists of two parts: atoms and laser lights, which are the main contributing pieces to extract the total interferometric phase. Since the phase is a scalar, it does not depend on the selection of the coordinate system, and we may equivalently observe this interacting atom–laser system in different coordinate systems. In the freely falling system attached to the atoms, all relativistic effects are reflected in the laser lights, while in the laser-platform coordinate system, i.e., the laboratory system, all relativistic effects are reflected in the atoms. To a large extent, we apply here the latter idea.

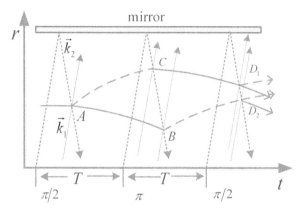

Fig. 1: Spacetime diagram of light-pulse atom interferometry. The solid and dashed lines represent the motions of the atoms in the ground and excited states, respectively, and the dotted lines depict the laser beams manipulating the atoms.

For a typical atomic gravimeter, one of the two Raman beams is reflected by a mirror (see $\vec{k}_2$ beam in Fig. 1). As the $\vec{k}_2$ beam reaches the atom later than the $\vec{k}_1$ beam, and stimulated Raman transitions occur only when both of the two beams interact with atoms, the $\vec{k}_1$ beam can be considered as "background light," and the $\vec{k}_2$ beam can be considered as "control light." Based on Fig. 1, the total interferometric phase shift can be calculated, which contains three parts: the atomic-propagation phase shift, the laser phase shift, and the separation phase shift. Conventionally, to calculate the relativistic phase shift, one should first solve the geodesic equations of atoms and photons in the gravitational field to determine their trajectories, next treat the five intersections A, B, C, D_1, and D_2, then calculate the propagation phase shift with the path integral method, and finally obtain the total interferometric phase shift as well as the gravitational acceleration. However, as the integral intervals for the two paths are temporally different due to the FSL, the calculation is not straightforward and usually involves a computer-based approach to give a scalar expression. The main idea of this work is to perform a coordinate transformation to transfer FSL effects of the "control light" to the atomic lagrangian, chosen such that the integral intervals of the classical lagrangians for the two paths are temporally identical in the new coordinate system. The velocity of the $\vec{k}_2$ light becomes $dr'/dt' = \infty$, and the atomic lagrangian contains the FSL disturbance in the new system. Then, the total phase shift can be calculated using the Bordé ABCD matrix and perturbation methods.

3. Results

We calculated the FSL effect in Special Relativity, and derived a more complete vectorial expression.[7,8] We find that, except for the results given by Kasevich's group[5,6] and Chu's group,[2,3] the FSL correction also includes $-2\frac{\vec{v}(T)\cdot\vec{e}_k}{c}\frac{\alpha_1-\alpha_2}{\vec{k}_{\text{eff}}\cdot\vec{g}_0}$, which is of the same order of magnitude as the other terms, but had been missed before. We further make clear the physical roots of these corrective terms. The main interferometric phase shift arises from the atom absorbing or emitting laser phase shifts, which can be simply described by

$$\phi_{\text{laser}} = \vec{k}_{\text{eff}} \cdot \vec{r} - \omega_{\text{eff}}t + \phi_0 \rightarrow \left(\frac{\omega_1\vec{n}_1 - \omega_2\vec{n}_2}{c} + \frac{\alpha_1\vec{n}_1 - \alpha_2\vec{n}_2}{c}T \right) \cdot \vec{r}(T, \delta T)$$
$$- [\omega_1 - \omega_2 + (\alpha_1 - \alpha_2)T]\, t(T, \delta T) + \phi_0. \quad (1)$$

Here, δT is the time delay due to the finite propagation speed of light, and α_1 and α_2 are the frequency chirps, which should be introduced to compensate the Doppler shift due to atomic motion. We defined the $1/c$ terms as related to the FSL effect. Since δT is related to $1/c$, the FSL effect includes three parts: the pure FSL time delay, the coupling of the frequency chirp and the time delay, and the chirp-dependent changes of the wave vector. In addition, we find the FSL correction depends on the propagating directions of the light beams involved in the measurement process. Therefore, the subterms of the FSL correction may be experimentally tested by adjusting the experimental configuration. That is why Cheng et al.[4] reported they only verified experimentally the FSL effect associated with the coupling of the frequency chirp and the time delay.

Based on the calculation for the FSL correction, we derived a more complete relativistic expression of the interferometric phase shift, which is suitable to analyze the atoms' motion in three dimensions. In addition, this result has been the first to take the relativistic effects in atomic gravimeters due to the Earth's rotation into account, and also completed the analysis of effects related to gravity gradients.

4. Conclusion and prospects

This work has developed an analytical method for modeling atom interferometers that clearly exposes the FSL effect and provides a more complete relativistic description of the measurement process. These results are important for testing GR with atomic interferometry. In the near future, we will, on the one hand, consider exploring error-elimination schemes, such as the frequency-shift gravity-gradient compensation technique,[10] a work already in progress;[11] on the other hand, we want to explore GR test schemes, such as searches for Lorentz violation and gravitational waves.

Acknowledgments

This work is supported by the Postdoctoral Science Foundation of China under grant Nos. 2017M620308 and 2018T110750.

References

1. S. Wajima *et al.*, Phys. Rev. D **55**, 1997 (1964).
2. A. Peters, Ph.D. thesis, Stanford University (1998).
3. A. Peters, K.Y. Chung, and S. Chu, Metrologia **38**, 25 (2001).
4. B. Cheng, P. Gillot, S. Merlet, and F. Pereira Dos Santos, Phys. Rev. A **92**, 063617 (2015).
5. S. Dimopoulos, P.W. Graham, J.M. Hogan, and M.A. Kasevich, Phys. Rev. Lett. **98**, 111102 (2007).
6. S. Dimopoulos, P.W. Graham, J.M. Hogan, and M.A. Kasevich, Phys. Rev. D **78**, 042003 (2008).
7. Y.J. Tan, C.G. Shao, and Z.K. Hu, Phys. Rev. A **94**, 013612 (2016).
8. Y.J. Tan, C.G. Shao, and Z.K. Hu, Phys. Rev. A **96**, 023604 (2017).
9. Y.J. Tan, C.G. Shao, and Z.K. Hu, Phys. Rev. D **95**, 024002 (2017).
10. A. Roura, Phys. Rev. Lett. **118**, 160401 (2017).
11. Y.J. Wang, X.Y. Lu, Y.J. Tan, C.G. Shao, and Z.K. Hu, Phys. Rev. A **98**, 053604 (2018).

Lorentz Violation and Riemann–Finsler Geometry

Benjamin R. Edwards

Physics Department, Indiana University, Bloomington, IN 47405, USA

The general charge-conserving effective scalar field theory incorporating violations of Lorentz symmetry is presented. The dispersion relation is used to infer the effect of spin-independent Lorentz violation on point-particle motion. A large class of associated Finsler spaces is derived, and the properties of these spaces are explored.

1. Introduction

Connections between Riemann–Finsler spaces and theories with Lorentz violation have recently been uncovered.[1] A lack of physical examples is an obstacle on the path toward developing a strong intuition about Finsler spaces. In the first section, the general effective quadratic scalar field theory incorporating violations of Lorentz symmetry will be developed. In the next section, a method to generate the lagrangian describing the motion of an analogue point particle experiencing spin-independent Lorentz violation is derived. The last section explores the properties of these Finsler spaces. These proceedings are based on results in Ref. 2.

2. Field theory

For a complex scalar field ϕ propagating in an n-dimensional Minkowski spacetime with metric $\eta_{\mu\nu}$, the quadratic Lagrange density incorporating Lorentz violation is

$$\mathcal{L}(\phi, \phi^\dagger) = \partial^\mu \phi^\dagger \partial_\mu \phi - m^2 \phi^\dagger \phi + \tfrac{1}{2} \big[\partial_\mu \phi^\dagger (\widehat{k}_c)^{\mu\nu} \partial_\nu \phi - i\phi^\dagger (\widehat{k}_a)^\mu \partial_\mu \phi + \text{h.c.} \big]. \quad (1)$$

The Lorentz violation is realized by the CPT-odd operator $(\widehat{k}_a)^\mu$ and the CPT-even operator $(\widehat{k}_c)^{\mu\nu}$, each of which can include coefficients for Lorentz violation associated with operators of arbitrarily large mass dimension d. Extensions of this theory may be used to describe, for example, CPT violation in neutral-meson oscillations.[3]

Field redefinitions can eliminate any traces present in the coefficients for Lorentz violation by absorbing them into the terms with lower mass dimension. We can therefore take them to be traceless without loss of generality. The commutativity of derivatives implies that they are totally symmetric in all their indices. From these considerations, it is found that the coefficients contain $(2d - n + 2)(d - 1)!/(d - n + 2)(n - 2)!$ independent components. The hermiticity of $\mathcal{L}$ implies these operators are themselves hermitian. In the special case of hermitian scalar fields, the term involving $(\widehat{k}_a)^\mu$ is proportional to a total derivative. It follows that CPT symmetry is guaranteed when $\phi = \phi^\dagger$.

The dispersion relation for this theory is found to be

$$p^2 - m^2 + (\widehat{k}_c)^{\mu\nu} p_\mu p_\nu - (\widehat{k}_a)^\mu p_\mu = 0, \tag{2}$$

where the operators $(\widehat{k}_c)^{\mu\nu}$ and $(\widehat{k}_a)^\mu$ are expressed in momentum space as

$$(\widehat{k}_c)^{\mu\nu} = \sum_{d=n}^{\infty} (k_c^{(d)})^{\mu\nu\alpha_1\alpha_2\cdots\alpha_{d-n}} p_{\alpha_1} p_{\alpha_2} \cdots p_{\alpha_{d-n}},$$

$$(\widehat{k}_a)^\mu = \sum_{d=n-1}^{\infty} (k_a^{(d)})^{\mu\alpha_1\cdots\alpha_{d-n+1}} p_{\alpha_1} p_{\alpha_2} \cdots p_{\alpha_{d-n+1}}, \tag{3}$$

with the sums running over even powers of p. For brevity, both types of coefficients will be expressed without the a or c subscripts in what follows, and the appropriate sign difference will be absorbed into the $k^{(d)}$ coefficient where the CPT properties will be determined by the mass dimension d.

3. Classical kinematics

A method has been developed to extract point-particle lagrangians from a given field theory.[4] Using the three equations

$$R(p) = 0, \tag{4}$$

$$\frac{\partial p_0}{\partial P_j} = -\frac{u^j}{u^0}, \tag{5}$$

$$L = -u^\mu p_\mu, \tag{6}$$

the idea is to identify the centroid of a localized wave packet with the point particle. The method starts with the dispersion relation $R(p)$. Next, the components of the classical velocity u^μ of the particle are related to the group velocity of the corresponding wave packet. The last equation is required by translation invariance of L, along with the requirement that L be one-homogeneous in the velocity, $L(\lambda u) = \lambda L(u)$. The first two relations can then be used to eliminate the components of the n-momentum to write L only as a function of the velocity u^μ.

These equations combine to produce a quadratic polynomial in L. For the case $d = n$, solving this quadratic leads to the exact lagrangian. For the nonminimal cases $d \geq 5$, corrections to the usual lagrangian are determined through an iterative procedure leading to the third-order correction

$$
\begin{aligned}
L_3^{(d)} = L_0^{(d)} \Big[& 1 - \tfrac{1}{2}\widetilde{k}^{(d)} - \tfrac{1}{8}(d - n + 1)^2(\widetilde{k}^{(d)})^2 \\
& + \tfrac{1}{8}(d - n + 2)^2 \, \widetilde{k}^{(d)}{}_\alpha \, \widetilde{k}^{(d)\alpha} - \tfrac{1}{16}(d - n + 1)^4(\widetilde{k}^{(d)})^3 \\
& + \tfrac{1}{16}(d - n + 1)(d - n + 2)^2(2d - 2n + 1) \, \widetilde{k}^{(d)} \, \widetilde{k}^{(d)}{}_\alpha \, \widetilde{k}^{(d)\alpha} \\
& - \tfrac{1}{16}(d - n + 1)(d - n + 2)^3 \, \widetilde{k}^{(d)}{}_\alpha \, \widetilde{k}^{(d)\alpha\beta} \, \widetilde{k}^{(d)}{}_\beta \Big],
\end{aligned}
\tag{7}
$$

where the

$$
\widetilde{k}^{(d)} = m^{n-d}(k^{(d)})_{\alpha_1\alpha_2\cdots\alpha_{d-n+2}} \hat{u}^{\alpha_1}\hat{u}^{\alpha_2}\cdots\hat{u}^{\alpha_{d-n+2}},
$$

$$
\widetilde{k}^{(d)}{}_{\alpha_1\cdots\alpha_l} = m^{n-d}(k^{(d)})_{\alpha_1\cdots\alpha_l\alpha_{l+1}\cdots\alpha_{d-n+2}} \hat{u}^{\alpha_{l+1}}\cdots\hat{u}^{\alpha_{d-n+2}},
\tag{8}
$$

contain the directional dependence $\hat{u}^\mu = u^\mu/u$, $u = \sqrt{u^\mu\eta_{\mu\nu}u^\nu}$. This lagrangian matches the first-order correction found by Reis and Schreck for the nonminimal fermion sector using an ansatz-based technique.[5]

4. Finsler geometry

The Finsler structure (or Finsler norm) plays a central role in the study of Finsler spaces. Classical lagrangians satisfy many of the requirements in the definition of Finsler structures. Though there are important differences preventing the lagrangians derived above from fulfilling the definition of a Finsler structure, a method exists to generate associated Finsler structures from a given lagrangian.[6] For the lagrangians developed in the previous section, the prescription to generate a Finsler structure in this case is given by $p_x(u) \rightarrow (-i)^n p_x(y)$, $k^{(d)x} \rightarrow i^n k^{(d)x}$, $L \rightarrow -F = -y \cdot p$, $u^x \rightarrow (i)^n y^x$. For example, the first-order limit of the lagrangian given in Eq. (7), leads to the associated Finsler structure

$$
F^{(d)} = y - \tfrac{1}{2}y\widetilde{k}^{(d)}.
\tag{9}
$$

The fundamental tensor of a Finsler space determines the metric and therefore also the geodesics. The definition of this tensor is $g_{jk} = \tfrac{1}{2}\partial^2 F^2/\partial y^j \partial y^k$. For the limit under consideration, the fundamental tensor is given by

$$
\begin{aligned}
g_{jk}^{(d)} = \; & r_{jk}(1 + \tfrac{1}{2}(d - n)\widetilde{k}^{(d)}) - \tfrac{1}{2}(d - n + 2)(d - n + 1)\widetilde{k}^{(d)}{}_{jk} \\
& + \tfrac{1}{2}(d - n)(d - n + 2)[\widetilde{k}^{(d)}{}_j\hat{y}_k + \widetilde{k}^{(d)}{}_k\hat{y}_j - \widetilde{k}^{(d)}\hat{y}_j\hat{y}_k].
\end{aligned}
\tag{10}
$$

Inspection shows $g^{(d)}_{jk}$ reduces to a purely riemannian one for the cases $d = n$ and $d = n - 2$. This is consistent with the fact that the $d = n$ coefficient can be absorbed into the metric at the level of the field theory, while a $d = n - 2$ coefficient would correspond to a rescaling of the mass term.

The situation is not as straightforward for other values of mass dimension. It has been demonstrated that the nonvanishing of the Cartan torsion implies non-riemmannian nature of the underlying space.[7] Calculation of this tensor shows it vanishes for $d = n$ and $d = n - 2$, and also in the case of $n = 1$, which represents a Riemann curve, but is nonvanishing in other cases. Calculation of the Matsumoto torsion[8] shows only $d = n - 1$ reduces to a Randers metric.

The geodesics are obtained from the geodesic equation $F \frac{d}{d\lambda}(y^j/F) + G^j = 0$. A calculation shows the geodesic spray coefficients G^j are

$$\frac{1}{y^2} G^j = \frac{1}{2} \widetilde{D}^j \widetilde{k}^{(d)} + \frac{1}{2}(d - n)\hat{y}^j \widetilde{D}_\bullet \widetilde{k}^{(d)}$$
$$- \frac{1}{2}(d - n + 2)r^{jl} \widetilde{D}_\bullet \widetilde{k}^{(d)}{}_l + \widetilde{\gamma}^j{}_{\bullet\bullet}, \tag{11}$$

where a $\bullet$ subscript denotes contraction of $\hat{y}^j$ with a lower j index, with all contractions exterior to any derivatives.

It is clear from this expression that if the background field is covariantly constant with respect to the riemannian metric, $\widetilde{D}_j \widetilde{k}^{(d)}{}_l = 0$, then the geodesics are unaffected. This situation was conjectured to hold in general and is confirmed here, but remains unproved.

Acknowledgments

This work was supported in part by the US Department of Energy under grant No. DE-SC0010120 and by the Indiana University Center for Spacetime Symmetries.

References

1. V.A. Kostelecký, Phys. Rev. D **69**, 105009 (2004).
2. B.R. Edwards and V.A. Kostelecký, Phys. Lett. B **786**, 319 (2018).
3. B.R. Edwards and V.A. Kostelecký, in preparation.
4. V.A. Kostelecký and N. Russell, Phys. Lett. B **693**, 443 (2010).
5. J.A.A.S. Reis and M. Schreck, Phys. Rev. D **97**, 065019 (2018).
6. V.A. Kostelecký, Phys. Lett. B **701**, 137 (2011).
7. A. Deicke, Arch. Math. **4**, 45 (1953).
8. M. Matsumoto, Tensor **24**, 29 (1972); M. Matsumoto and S. Hōjō, Tensor **32**, 225 (1978).

Physics Beyond the Standard Model with DUNE: Prospects for Exploring Lorentz and CPT Violation

Célio A. Moura

Centro de Ciências Naturais e Humanas, Federal do ABC,
Avenida dos Estados, 5001, Santo André - SP, 09210-580, Brazil

On behalf of the DUNE Collaboration

In this talk, we present the physics potential of DUNE in the context of neutrino oscillations, nonstandard interactions, and Lorentz and CPT and violation. If the DUNE data are consistent with the standard oscillations for three massive neutrinos, nonstandard neutrino-interaction effects on the order of a tenth of G_F can be ruled out in the propagation. DUNE can also help to improve the present bounds on Lorentz and CPT violation by several orders of magnitude contributing as a very important experiment to test quantum field theory.

DUNE[1] will measure neutrino oscillations over a 1300 km baseline from Fermilab, near Chicago, to SURF in South Dakota. It can improve the present limits on Lorentz and CPT violation by several orders of magnitude.[2–4] One of the predictions of CPT invariance is that particles and antiparticles have the same masses. Using neutrino-oscillation data, we can compare the mass splittings and mixing angles of neutrinos with those of antineutrinos.[3] Any differences in the neutrino and antineutrino spectrum imply the violation of CPT symmetry.

In Ref. 3, the authors derived the most up-to-date bounds on CPT invariance from the neutrino sector. The data used to derive these bounds are the same as in the global fit to neutrino oscillations in Ref. 5. See Ref. 3 and references therein for the complete data set used, as well as the parameters to which they are sensitive. At the moment it is not possible to set any bound on $|\delta - \overline{\delta}|$, since all possible values of δ or $\overline{\delta}$ are allowed by the data. The preferred intervals of δ obtained in Ref. 5 can only be obtained after combining the neutrino and antineutrino data samples. The limits on $\Delta(\Delta m_{31}^2)$ and $\Delta(\Delta m_{21}^2)$ are already better than the ones derived from the neutral-kaon system[6] and should be regarded as the best bounds on CPT violation on the mass squared so far. See Table 1 for these results.

Table 1: Oscillation parameters used in this work and CPT-violation bounds. All mass-squared values are given in units of eV^2.

Parameter	Value	Neutrino CPTv Bounds [3]	$K^0\bar{K}^0$ Bound [3,6]
Δm_{21}^2	7.56×10^{-5}	$\|\Delta m_{21}^2 - \Delta\overline{m}_{21}^2\| < 4.7 \times 10^{-5}$	
Δm_{31}^2	2.55×10^{-3}	$\|\Delta m_{31}^2 - \Delta\overline{m}_{31}^2\| < 0.37 \times 10^{-3}$	
$\sin^2\theta_{23}$	0.43, 0.50, 0.60	$\|\sin^2\theta_{23} - \sin^2\overline{\theta}_{23}\| < 0.32$	
$\sin^2\theta_{13}$	0.02155	$\|\sin^2\theta_{13} - \sin^2\overline{\theta}_{13}\| < 0.03$	
$\sin^2\theta_{12}$	0.321	$\|\sin^2\theta_{12} - \sin^2\overline{\theta}_{12}\| < 0.14$	
δ	1.50π		$\|\Delta m_{K^0\bar{K}^0}^2\| < 0.25$

To study the sensitivity of the DUNE experiment to CPT violation by analyzing neutrino and antineutrino oscillation parameters separately, as in Ref. 3, we assume the neutrino oscillations are parameterized by the usual PMNS matrix U_{PMNS} with parameters θ_{12}, θ_{13}, θ_{23}, Δm_{21}^2, Δm_{31}^2, and δ, while the antineutrino oscillations are parameterized by a matrix $\overline{U}_{\text{PMNS}}$ with parameters $\overline{\theta}_{12}$, $\overline{\theta}_{13}$, $\overline{\theta}_{23}$, $\Delta\overline{m}_{21}^2$, $\Delta\overline{m}_{31}^2$, and $\overline{\delta}$, with the same probability function. To simulate the future neutrino data signal at DUNE, we assume the true values for neutrinos and antineutrinos to be the ones in Table 1. In the statistical analysis, we vary freely all the oscillation parameters, except the solar ones, which are fixed to their best-fit values throughout the simulations. Given the great precision in the determination of the reactor mixing angle by short-baseline reactor experiments, in this analysis, we use a prior on $\overline{\theta}_{13}$, but not on θ_{13}. We also consider three different values for the atmospheric angles, as indicated in Table 1.

To test the sensitivity at DUNE we perform the simulations assuming $\Delta x = |x - \overline{x}| = 0$, where x is any of the oscillation parameters. Then we estimate the sensitivity to $\Delta x \neq 0$. To do so we calculate two χ^2 grids, one for neutrinos and another one for antineutrinos, varying the four parameters of interest. After minimizing over all parameters except x and $\overline{x}$, we calculate

$$\chi^2(\Delta x) = \chi^2(|x - \overline{x}|) = \chi^2(x) + \chi^2(\overline{x}). \tag{1}$$

The results are presented in Fig. 1. The three different lines, labeled as "high," "max," and "low," refer to the assumed value for the atmospheric angle: in the lower octant (low), maximal mixing (max), or in the upper octant (high). There is neither sensitivity to $\Delta(\sin^2\theta_{13})$, where the 3σ bound would be of the same order as the current measured value for $\sin^2\overline{\theta}_{13}$, nor to $\Delta\delta$, where no single value of the parameter would be excluded at more than

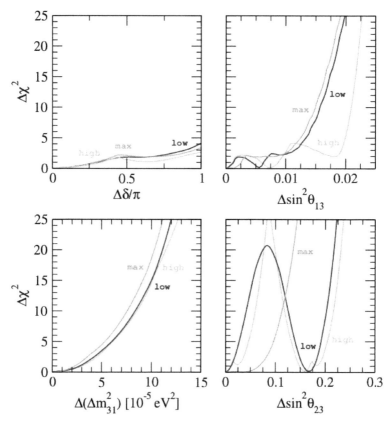

Fig. 1: The sensitivities of DUNE to the difference in neutrino and antineutrino parameters: $\Delta\delta$, $\Delta(\Delta m_{31}^2)$, $\Delta(\sin^2\theta_{13})$, and $\Delta(\sin^2\theta_{23})$ for the atmospheric angle in the lower octant, in the upper octant, and for maximal mixing.[3]

2σ. Interesting results are obtained for $\Delta(\Delta m_{31}^2)$ and $\Delta(\sin^2\theta_{23})$. DUNE can put stronger bounds on the difference of the atmospheric mass splittings, namely $\Delta(\Delta m_{31}^2) < 8.1 \times 10^{-5}\,\mathrm{eV}^2$, improving the current neutrino bound by one order of magnitude. For the atmospheric angle, we obtain different results depending on the true value assumed in the simulation of DUNE data. In the lower right panel of Fig. 1, we see the different behavior obtained for θ_{23} lying in the lower octant, being maximal, and lying in the upper octant. As one might expect, the sensitivity increases with $\Delta\sin^2\theta_{23}$ in the case of maximal mixing. However, if the true value lies in the lower or upper octant, a degenerate solution appears in the complementary octant.

In different types of neutrino oscillation experiments, as for example accelerators, neutrino and antineutrino data are obtained in separate experimental runs. However, the usual procedure followed by the experimental collaborations, as well as the global oscillation fits, as, e.g., in Ref. 5, assume CPT invariance and analyze the full data sample in a joint way. If CPT is violated in nature, the outcome of the joint data analysis might give rise to a solution that results from the combined analysis but does not correspond to the true solution of any channel. Under the assumption of CPT conservation, the χ^2 functions are computed according to $\chi^2_{\text{total}} = \chi^2(\nu) + \chi^2(\overline{\nu})$, and assuming that the same parameters describe both neutrino and antineutrino flavor oscillations. In contrast, in Eq. (1) we first profiled over the parameters in neutrino and antineutrino mode separately and then added the profiles.

To conclude, DUNE will measure neutrino oscillations with high precision and can test to an unprecedented extent the assumptions of the CPT theorem, such as Lorentz invariance.[4]

Acknowledgments

CAM thanks FAPESP grant No. 2014/19164-6, UFABC, and the CPT'19 organizers.

References

1. DUNE Collaboration, B. Abi *et al.*, *The DUNE Far Detector Interim Design Report Volume 1: Physics, Technology and Strategies*, arXiv:1807.10334.
2. R.F. Streater and A.S. Wightman, *PCT, spin and statistics, and all that*, Addison-Wesley, Redwood City, 1989; G. Barenboim and J.D. Lykken, Phys. Lett. B **554**, 73 (2003); V.A. Kostelecký and M. Mewes, Phys. Rev. D **69**, 016005 (2004); J.S. Díaz, V.A. Kostelecký, and M. Mewes, Phys. Rev. D **80**, 076007 (2009); V.A. Kostelecký and M. Mewes, Phys. Rev. D **85**, 096005 (2012).
3. G. Barenboim, C.A. Ternes, and M. Tórtola, Phys. Lett. B **780**, 631 (2018).
4. G. Barenboim, M. Masud, C.A. Ternes, and M. Tórtola, Phys. Lett. B **788**, 308 (2019).
5. P.F. de Salas *et al.*, *Status of neutrino oscillations 2017*, arXiv:1708.01186.
6. B. Schwingenheuer *et al.*, Phys. Rev. Lett. **74**, 4376 (1995).

Lorentz Violation in High-Energy Hadron Collisions

E. Lunghi

Physics Department, Indiana University, Bloomington, IN 47405, USA

We present an overview of Lorentz violation in high-energy collisions focusing on deep inelastic scattering and Drell–Yan at hadron-hadron colliders. We further discuss the bounds that can be attainable by sidereal-time studies of these processes at HERA, the planned Electron–Ion Collider, and the LHC.

Invariance of physical laws under Lorentz transformations is the most tested symmetry in Nature. Lorentz invariance constitutes the foundation of modern physics to such an extent that formulating consistent theories in which this symmetry is broken is a severe challenge. The Standard-Model Extension[1–3] (SME) provides a model-independent framework in which Lorentz-violating effects can be described. The foundation of the SME lies in the preservation of invariance under observer Lorentz transformations. Every term in the SME lagrangian is built out of standard building blocks (scalar, spinor, and tensor fields, coupling constants, and masses) and of coefficients for Lorentz violation, which transform as tensors under observer transformations but are constant under particle Lorentz transformations. This set up can be thought of as the result of a spontaneous breaking of Lorentz symmetry, though this point of view is not necessary.

Bounds on Lorentz-violating effects in stable or long-lived particles (e.g., e, μ, γ, p, n) are extremely strong. The corresponding coefficients (which are fundamental parameters for the leptons and the photon, and effective parameters for the nucleons) are very tightly constrained.[4] On the other hand, constraints on coefficients in the quark sector are very weak due to the difficulty in relating quark and hadron properties (for attempts along these lines, see Refs. 5,6).

In a series of recent papers[7–10] we showed how to connect directly quark coefficients to observables in collisions at high energies. The presence of these coefficients induces a sidereal time variation. The irreducible background to this modulation is controlled by the fraction of the total experimental uncertainty which is uncorrelated with respect to time binning. The latter is usually statistical in origin. For this reason, we focused on processes for which statistical errors are the smallest. Two prominent examples are deep inelastic scattering (DIS) at electron–proton colliders and Drell–Yan di-lepton pair production at hadron colliders. We refer to Ref. 10 for a detailed proof of factorization theorems for DIS and Drell–Yan within the SME. In the following we summarize the main results.

The SME coefficients that we consider are:

$$\mathcal{L} = \sum_{f=u,d} \left[\tfrac{1}{2} \bar{\psi}_f (\eta^{\mu\nu} + c_f^{\mu\nu}) \gamma_\mu i D_\nu \psi_f - a_f^{(5)\mu\alpha\beta} \bar{\psi}_f \gamma_\mu i D_{(\alpha} i D_{\beta)} \psi_f + \text{h.c.} \right],$$

(1)

where $c_f^{\mu\nu}$ are dimensionless coefficients, which belong to the so-called minimal SME, and $a_f^{(5)\mu\alpha\beta}$ are negative-dimension nonrenormalizable coefficients. The latter are interesting because their effects tend to be enhanced in very high energy interactions.

The prescription to calculate a generic lepton–hadron DIS cross section is relativelty simple.[10] First, one needs to calculate the matrix element squared for the quark-level transition and perform the phase-space integration. In order to avoid complications associated with final-state SME particles, these steps are best achieved by employing the optical theorem: the integrated rate is obtained from the imaginary part of the forward lepton–hadron amplitude calculated using the standard Cutkosky rules. The final cross section is then obtained by multiplying by the proton flux factor (which depends only on the proton SME coefficients, which are experimentally tiny[4]) and convoluting with the quark parton distribution functions (PDFs). An explicit expression for the latter, valid in the presence of the minimal coefficients $c_f^{\mu\nu}$, is:

$$f_f(\xi, c_f^{pp}) = \int \frac{d\lambda}{2\pi} e^{-i\xi p \cdot n\lambda} \langle p | \bar{\psi}(\lambda \tilde{n}_f) \frac{\not{n}}{2} \psi(0) | p \rangle,$$

(2)

where n and $\bar{n}$ are two lightcone vectors determined by the proton momentum $p^\mu = \bar{n}^\mu (n \cdot p)$, and $\tilde{n}_f^\mu = (\eta^{\mu\nu} + c_f^{\mu\nu}) n_\nu$. This expression reduces to the standard definition of the PDFs in the $c_f \to 0$ limit. More importantly, it is possible to show that the nth moment of this PDF is in one-to-one correspondence with the matrix element of the operator of dimension $n+3$ that appears in the Operator Product Expansion proof of the factorization theorem for DIS.[4] A similar result can be obtained for the nonminimal $a_f^{(5)}$ coefficients. In particular, this shows that the PDFs can only depend on contractions of the coefficients for Lorentz violation with the proton

Table 1: Bounds on selected u-quark coefficients at HERA, the EIC, and the LHC. Bounds are reported in units of 10^{-5} and $10^{-5}\,\mathrm{GeV}^{-1}$ for the c_f and $a_f^{(5)}$ coefficients, respectively.

	HERA	EIC	LHC
$\lvert c_u^{XZ} \rvert$	63	0.23	7.3
$\lvert c_u^{YZ} \rvert$	65	0.23	7.1
$\lvert c_u^{XY} \rvert$	31	0.26	2.7
$\lvert c_u^{XX} - c_u^{YY} \rvert$	98	0.74	15
$\lvert a_u^{(5)TXZ} \rvert$	1.3	0.013	0.0031
$\lvert a_u^{(5)TYZ} \rvert$	1.3	0.013	0.0031
$\lvert a_u^{(5)TXY} \rvert$	0.65	0.036	0.0014
$\lvert a_u^{(5)TXX} - a_u^{(5)TYY} \rvert$	3.1	0.045	0.0064

momentum: c_f^{pp} for the minimal case and $a_f^{(5)ppp}$, $a_f^{(5)p\mu}{}_\mu$, and $a_f^{(5)\mu p}{}_\mu$ for the nonminimal one. The dependence of the PDFs on the coefficients is nonperturbative and will be investigated in a forthcoming publication.

The calculation of the Drell–Yan cross section is very similar: the partonic squared matrix element calculated in the SME is convoluted with the *same* PDFs introduced above. The only additional complication is that we need to take into account that the di-lepton momentum is connected to the momenta of two partons which are on-shell in the SME: this leads to the appearance of an extra correction term in the final cross section.[10]

All coefficients for Lorentz violation are defined in the Sun-centered celestial equatorial frame; therefore, in the laboratory frame most coefficients induce a sidereal time variation that depends on the location and orientation of the experiment.

We focus on DIS at HERA[11] and at the planned Electron–Ion Collider (EIC)[12] and on Drell–Yan at CMS.[13] The expected bounds on the coefficients c_f and $a_f^{(5)}$ can be calculated by taking existing (or Monte Carlo generated) data and simulating the sidereal time binning by carefully considering what fraction of the experimental uncertainty is correlated in time (see Refs. 7,8 for details).

In Table 1, we present bounds on selected coefficients that can be constrained by both DIS and Drell–Yan. We first note that expected bounds from EIC measurements are expected to place bounds that are more than an order of magnitude stronger than those from HERA. This is due to the much larger luminosity of the EIC. More important is the comparative advantage that LHC measurements have in the extraction of bounds on the coefficients $a_f^{(5)}$: while bounds from Drell–Yan on $c_f^{\mu\nu}$ are weaker than those attainable at the EIC, bounds on the nonminimal coefficients $a_f^{(5)}$ are an order of magnitude stronger than the EIC ones. The reason for this behavior is that the $a_f^{(5)}$ have negative mass dimension implying that in physical observables they appear multiplied by the typical experimental energies, which are much larger at the LHC. We conclude that both the EIC and the LHC have the potential to place strong constraints on various quark-level coefficients for Lorentz violation.

References

1. D. Colladay and V.A. Kostelecký, Phys. Rev. D **55**, 6760 (1997).
2. D. Colladay and V.A. Kostelecký, Phys. Rev. D **58**, 116002 (1998).
3. V.A. Kostelecký, Phys. Rev. D **69**, 105009 (2004).
4. *Data Tables for Lorentz and CPT Violation,* V.A. Kostelecký and N. Russell, 2019 edition, arXiv:0801.0287v12.
5. R. Kamand, B. Altschul, and M.R. Schindler, Phys. Rev. D **95**, 056005 (2017).
6. B. Altschul and M.R. Schindler, arXiv:1907.02490.
7. V.A. Kostelecký, E. Lunghi, and A.R. Vieira, Phys. Lett. B **769**, 272 (2017).
8. E. Lunghi and N. Sherrill, Phys. Rev. D **98**, 115018 (2018).
9. V.A. Kostelecký and Z. Li, Phys. Rev. D **99**, 056016 (2019).
10. V.A. Kostelecký, E. Lunghi, N. Sherrill, and A.R. Vieira, arXiv:1911.04002.
11. H1 and ZEUS Collaborations, H. Abramowicz *et al.*, Eur. Phys. J. C **75**, 580 (2015).
12. A. Accardi *et al.*, Eur. Phys. J. A **52**, 268 (2016).
13. CMS Collaboration, A.M. Sirunyan *et al.*, arXiv:1812.10529.

Explicit Spacetime-Symmetry Breaking in Matter: The Reversed Vavilov–Čerenkov Radiation

O.J. Franca, L.F. Urrutia, and Omar Rodríguez-Tzompatzi

High Energy Physics Department, Universidad Nacional Autónoma de México, CDMX, 04510, Mexico

We show that reversed Vavilov–Čerenkov radiation occurs simultaneously with the forward output in naturally existing materials when an electric charge moves with a constant velocity perpendicular to the planar interface between two magnetoelectric media. Using the Green's function in the far-field approximation we calculate the angular distribution of the radiated energy per unit frequency obtaining a non-zero contribution in the backward direction.

1. Introduction

As a motivation for this work let us recall that spacetime-dependent couplings have been identified as possible sources of fundamental Lorentz-invariance violation.[1] Also, let us consider the photon sector of the Standard-Model Extension[2]

$$\mathcal{L} = -\tfrac{1}{4}F_{\mu\nu}F^{\mu\nu} - \tfrac{1}{4}(k_F)_{\kappa\lambda\mu\nu}F^{\kappa\lambda}F^{\mu\nu} + \tfrac{1}{2}(k_{AF})^{\kappa}\epsilon_{\kappa\lambda\mu\nu}A^{\lambda}F^{\mu\nu}. \qquad (1)$$

A term that is missing in the Lagrange density (1) is $(k_F)_{\kappa\lambda\mu\nu} = \theta\epsilon_{\kappa\lambda\mu\nu}$, with the axion coupling θ being a constant, simply because this contribution is a total derivative. Nevertheless, if we promote the constant θ to $\theta(x)$ we are able to describe the effective electromagnetic response of a host of interesting physical phenomena where spacetime symmetries are expected to be explicitly broken. In this work we deal with axion electrodynamics

$$\mathcal{L} = -\tfrac{1}{4}F_{\mu\nu}F^{\mu\nu} - \tfrac{\alpha}{32\pi^2}\,\vartheta(\mathbf{x})\epsilon_{\kappa\lambda\mu\nu}F^{\kappa\lambda}F^{\mu\nu}, \qquad (2)$$

which describes the electromagnetic response of magnetoelectric materials like topological insulators[3] and Weyl semimetals,[4] for example. Here we summarize the discovery of reversed Vavilov–Čerenkov radiation (RVCR) in naturally existing magnetoelectrics, reported in Ref. 5.

In 1968 Veselago[6] proposed that RVCR could be produced in materials characterized by negative permittivity ϵ and permeability μ. In this case a radiation cone is observed in the backward direction with respect to the velocity of the propagating charge. Since such materials are not found in nature, this proposal gave a major thrust to the design and construction of metamaterials that would provide the required properties in a given range of frequencies.[7]

2. RVCR in axion electrodynamics

We consider two semi-infinite media separated by the interface Σ at $z = 0$ with the axion field $\vartheta(z) = H(z)\vartheta_2 + H(-z)\vartheta_1$ and ϑ_1, ϑ_2 constants, where $H(z)$ is the Heaviside function. In order to suppress the transition radiation we take $\epsilon_1 = \epsilon_2 = \epsilon$ as a first approximation. We also set $\mu_1 = \mu_2 = 1$. The modified Maxwell equations resulting from the Lagrange density (2) are those corresponding to a normal material medium with constitutive relations $\mathbf{D} = \epsilon\mathbf{E} - \alpha\vartheta(z)\mathbf{B}/\pi$ and $\mathbf{H} = \mathbf{B} + \alpha\vartheta(z)\mathbf{E}/\pi$. These equations introduce field-dependent effective charge and current densities with support only at the interface Σ, which produce the magnetoelectric effect,[3] being governed by the parameter $\tilde{\theta} = \alpha(\vartheta_2 - \vartheta_1)/\pi$. To calculate the radiation fields we rewrite the modified Maxwell equations in terms of the potential A_μ ($c = 1$) and introduce the Green's function (GF) $G^\nu{}_\sigma(x, x')$ by setting $A^\mu(x) = \int d^4x' G^\mu{}_\nu(x, x') J^\nu(x')$, obtaining

$$\left([\Box^2]^\mu{}_\nu - \tilde{\theta}\delta(z)\varepsilon^{3\mu\alpha}{}_\nu\partial_\alpha\right) G^\nu{}_\sigma(x, x') = 4\pi\eta^\mu{}_\sigma\delta^4(x - x'). \tag{3}$$

Let us remark that the resulting θ-dependent contribution to the GF is a function of $(|z| + |z'|)$, instead of $|z - z'|$ as it is in the normal case. This is the mathematical origin of the RVCR. The required GF in the far-field approximation includes integrals of rapidly oscillating functions and its leading contribution is obtained by the stationary-phase approximation.

Now, let us consider a charge q moving at a constant velocity $v\hat{\mathbf{u}}$ with charge and current densities $\varrho(\mathbf{x}'; \omega) = \frac{q}{v}\delta(x')\delta(y')e^{i\omega\frac{z'}{v}}$, $\mathbf{j}(\mathbf{x}'; \omega) = \varrho v\hat{\mathbf{u}}$. Here, $\hat{\mathbf{u}}$ is the unit vector perpendicular to the interface, directed from region 1 to region 2. We henceforth assume $v > 0$ and consider the motion in the interval $z \in (-\zeta, \zeta)$, with $\zeta \gg v/\omega$. In the far-field approximation we calculate the electric field $\mathbf{E}$ starting from the potential A^μ and using the corresponding approximation of the GF obtained from Eq. (3). Then, we are able to calculate the spectral distribution (SD) of the radiation $d^2E/d\omega d\Omega = (\mathbf{E}^* \cdot \mathbf{E})\, nr^2/4\pi^2$ in the limit $r \to \infty$. The main point here is that the following integrals, which result from the convolution of the GF with the sources,

$$\mathcal{I}_1(\omega,\theta) = \int_{-\zeta}^{\zeta} dz' e^{i\frac{\omega z'}{v}(1-vn\cos\theta)} = \frac{2\sin(\zeta\Xi_-)}{\Xi_-}, \tag{4}$$

$$\mathcal{I}_2(\omega,\theta) = \int_{-\zeta}^{\zeta} dz' e^{in\omega|z'|\cos\theta + i\omega\frac{z'}{v}} = \frac{\sin(\zeta\tilde{\Xi}_-)}{\tilde{\Xi}_-} + 2i\frac{\sin^2(\frac{\zeta}{2}\tilde{\Xi}_-)}{\tilde{\Xi}_-}, \tag{5}$$

enter in the expression for the electric field. The notation is $\Xi_- = \frac{\omega}{v}(1 - vn\cos\theta)$ and $\tilde{\Xi}_- = \frac{\omega}{v}(1 - vn|\cos\theta|)$. More importantly, in the limit $\zeta \gg v/\omega$, $(\zeta \to \infty)$, the right-hand side of Eqs. (4) and (5) yields expressions like $\sin(\zeta\,\rho N)/(\rho N)$ which behave as $\pi\delta(\rho N)$. This delta-like behavior means that the contributions of the electric field to the SD are nonzero only when (i) $1 \pm vn\cos\theta = 0$ and/or (ii) $1 \pm vn|\cos\theta| = 0$. Following our conventions, the case (i) yields the standard Čerenkov condition $\cos\theta_C = 1/(nv)$. The second case (ii) opens up the possibility that $\cos\theta$ is negative yielding the reversed Čerenkov cone at $\theta_R = \pi - \theta_C$. In other words, the terms containing $\tilde{\Xi}_-$ make possible the production of radiation in the backward direction. The SD of such radiation is suppressed with respect to the forward output, but nevertheless it is different from zero.

To illustrate our results we choose medium 2 as the topological insulator TlBiSe$_2$, with $n_2 = 2$ prepared in such a way that $\tilde{\theta} = 11\alpha$, together with a normal insulator in medium 1 having $n_1 = n_2$. We consider the radiation emitted at the average frequency of $\omega = 2.5\,\text{eV}$ (500 nm) in the Čerenkov spectrum. The SD of the total radiation is shown in Fig. 1. In the left panel of the figure we plot a zoom in the backward direction showing the onset of the RVCR and making evident the high suppression of this new contribution with respect to the forward Vavilov–Čerenkov radiation. The details of the above calculations are reported in Ref. 5.

3. Summary

We have considered the radiation produced by an electric charge moving at a constant velocity $v\hat{\mathbf{u}}$ perpendicular to the interface $z = 0$ between two semi-infinite planar magnetoelectric media with the same refraction index. When v is higher than $1/n$ we discover the emission of RVCR codified in the angular distribution illustrated in Fig. 1. The main characteristics of this RVCR are: (i) The threshold condition $v > 1/n$ must be satisfied as in the standard case. (ii) The RVCR occurs for all frequencies in the Čerenkov spectrum and it is always accompanied by the forward Vavilov–Čerenkov radiation with the same frequency. (iii) The SD of the RVCR is suppressed with respect to the forward output by the factor $\tilde{\theta}^2/(8n^2)$.[5]

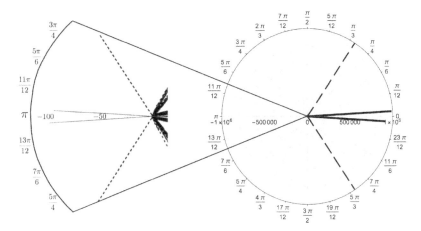

Fig. 1: Polar plot of the SD for the full Vavilov–Čerenkov radiation when $n_1 = n_2 = 2$, $\omega = 2.5\,\mathrm{eV}$, and $\tilde{\theta} = 11\alpha$, taken from Ref. 5. The dashed line corresponds to $v = 0.9$ and $\zeta = 343\,\mathrm{eV}^{-1}$, and the solid line to $v = 0.501$ and $\zeta = 4830\,\mathrm{eV}^{-1}$. The scale of the polar plot is arbitrary and runs from 0 to 10^6. The left panel in the figure is a zoom of the SD in the backward direction showing the onset of the RVCR. Here, the radial scale goes from zero to 10^2. The charge moves from left to right along the line $(\pi - 0)$.

Acknowledgments

OJF has been supported by the doctoral fellowship CONACYT-271523. OJF, LFU, and ORT acknowledge support from the projects: # 237503 from CONACYT and # IN103319 from Dirección General de Asuntos del Personal Académico (Universidad Nacional Autónoma de México). We thank VAK and collaborators for a wonderful meeting.

References

1. V.A. Kostelecký, R. Lehnert, and M.J. Perry, Phys. Rev. D **68**, 123511 (2003).
2. V.A. Kostelecký and M. Mewes, Phys. Rev. D **66**, 056005 (2002).
3. X.L. Qi, T.L. Hughes, and S. Ch. Zhang, Phys. Rev. B **78**, 195424 (2008); M.Z. Hasan and C.L. Kane, Rev. Mod. Phys. **82**, 3045 (2010).
4. N.P. Armitage, E.J. Mele, and A. Vishwanath, Rev. Mod. Phys. **90**, 015001 (2018).
5. O.J. Franca, L.F. Urrutia, and O. Rodríguez-Tzompantzi, Phys. Rev. D **99**, 116020 (2019).
6. V.G. Veselago, Soviet Physics Uspekhi **10**, 509 (1968).
7. Z. Duan *et al.*, Nature Communications **8**, 14901 (2017).

Searches for Lorentz-Violating Signals with Astrophysical Polarization Measurements

F. Kislat

Department of Physics and Astronomy and Space Science Center, University of New Hampshire, Durham, NH 03824, USA

Astrophysical observations are a powerful tool to constrain effects of Lorentz-invariance violation in the photon sector. Objects at high redshifts provide the longest possible baselines, and gamma-ray telescopes allow us to observe some of the highest energy photons. Observations include polarization measurements and time-of-flight measurements of transient or variable objects to constrain vacuum birefringence and dispersion. Observing multiple sources covering the entire sky allows the extraction of constraints on anisotropy. In this paper, I review methods and recent results on Lorentz- and CPT-invariance violation constraints derived from astrophysical polarization measurements in the framework of the Standard-Model Extension.

1. Introduction

Observer-independence of the speed of light postulated by Einstein's Theory of Special Relativity implies that the speed of light for a given observer is independent of frequency, direction of propagation, and polarization. Conversely, if Lorentz symmetry is broken, any of these may no longer hold, resulting in a vacuum that effectively acts like an anisotropic, birefringent, and/or dispersive medium. Theories of quantum gravity suggest that Lorentz invariance may be violated at the Planck scale, but effects must be strongly suppressed at attainable energies. Photons propagating over astrophysical distances enable some of the strongest tests as tiny effects will accumulate during propagation of the photons.

The Standard-Model Extension[1,2] (SME) is an effective-field-theory approach to describe low-energy effects of a more fundamental high-energy theory. It allows a categorization of effects, described in essence by a set of coefficients that are characterized in part by the mass dimension d of the corresponding operator. The SME photon dispersion relation can be written as

$$E = \left(1 - \varsigma^0 \pm \sqrt{(\varsigma^1)^2 + (\varsigma^2)^2 + (\varsigma^3)^2}\right) p, \tag{1}$$

where the ς^i can be written as an expansion in d and spherical harmonics. This results in $(d-1)^2$ independent coefficients $k^{(d)}_{(V)jm}$ for odd d describing anisotropy, dispersion, and birefringence; for even d, $2(d-1)^2 - 8$ independent birefringent coefficients $k^{(d)}_{(E)jm}$ and $k^{(d)}_{(B)jm}$ as well as $(d-1)^2$ nonbirefringent coefficients $c^{(d)}_{(I)jm}$ emerge. Effects of mass dimension d are typically assumed to be suppressed by M^{d-4}_{Planck}.

The individual coefficients $k^{(4)}_{(E)jm}$, $k^{(4)}_{(B)jm}$, $k^{(5)}_{(V)jm}$, and $c^{(6)}_{(I)jm}$ have been constrained using astrophysical polarization[3,4] and time-of-flight measurements,[5] and the $k^{(3)}_{(V)jm}$ have been constrained using CMB measurements.[6] At $d > 6$, linear combinations of coefficients have been constrained.[7] Here, we will review some recent results obtained from polarization measurements of astrophysical objects.

2. Birefringence

The polarization of an electromagnetic wave can be described by the Stokes vector $\boldsymbol{s} = (q, u, v)^T$, where q and u describe linear polarization and v describes circular polarization. Vacuum birefringence results in a change of the Stokes vector during the propagation of a photon given by[2]

$$\frac{d\boldsymbol{s}}{dt} = 2E\boldsymbol{\varsigma} \times \boldsymbol{s}, \tag{2}$$

which represents a rotation of $\boldsymbol{s}$ along a cone around the birefringence axis $\boldsymbol{\varsigma} = (\varsigma^1, \varsigma^2, \varsigma^3)^T$.

Most astrophysical emission mechanisms do not result in a strong circular polarization, but linear polarization can be fairly large reaching multiples of ten percent. In general, Eq. (2) results in a change of the linear polarization angle (PA) as the photon propagates, as well as linear polarization partially turning into circular polarization. However, SME operators of odd and even d result in distinctly different signatures.

If only odd-d coefficients are nonzero, $\varsigma^1 = \varsigma^2 = 0$ and $\boldsymbol{\varsigma}$ is oriented along the circular-polarization axis $\boldsymbol{v}$. In this case, linear polarization remains linear, but the PA makes full 180° rotations as $\boldsymbol{s}$ will rotate in the q–u plane. Any circular polarization would remain unaffected. If only even-d coefficients are nonzero, $\varsigma^3 = 0$ and $\boldsymbol{\varsigma}$ is in the q–u plane. Then, a Stokes vector with an initial $v = 0$ will in general rotate out of this plane acquiring a circular polarization, while at the same time the PA will undergo swings with a magnitude depending on the angle between $\boldsymbol{s}$ and $\boldsymbol{\varsigma}$. However, in this case linear polarization with a polarization angle of $\text{PA}_0 = \frac{1}{2}\arctan\left(-\varsigma^2/\varsigma^1\right)$ will remain unaffected.

For all $d \geq 4$, Eq. (2) results in an energy-dependent rate of change of s. Hence, birefringent Lorentz-violating effects can be constrained by measuring the (linear) polarization of astrophysical objects as a function of energy. Spectropolarimetric observations are ideally suited for this purpose as the change of PA with energy is measured directly resulting in the strongest constraints.

Constraints can also be derived from broadband measurements where polarization is integrated over the bandwidth of the instrument.[8] In this case, the rotation of the PA results in a net reduction of the observed polarization fraction (PF). Given a set of SME coefficients, the maximum observable PF, PF_{max}, can be calculated assuming the emission at the source is 100 % polarized. Any observed $PF > PF_{max}$ would then rule out this set of SME coefficients. Constraints obtained in this way tend to be significantly weaker than those obtained from spectropolarimetry.

Observations of a single source can be used to constrain linear combinations of SME coefficients. Typically, limits are obtained by restricting the analysis to a particular mass dimension d. In the even-d case, additional assumptions must be made due to the existence of the linearly polarized eigenmode of propagation. Observations of multiple objects can be combined to break all of these degeneracies and to obtain constraints on individual coefficients, even in the even-d case.

The strongest individual constraints result from hard x-ray polarization measurements of gamma-ray bursts (GRBs) profiting from the high photon energies and high redshifts. Typical constraints derived from these measurements are on the order of $10^{-34}\,\mathrm{GeV}^{-1}$ or better in the $d = 5$ case.[8,9] However, the number of bursts whose polarization has been measured is limited at this point and in particular the early results suffer from large systematic uncertainties. At least for $d < 6$, very strong constraints can also be obtained from optical polarization measurements of distant objects, such as active galactic nuclei and GRB afterglows. Systematic uncertainties of these measurements are smaller, and a much larger number of results is already available.

In two recent papers,[3,4] we have combined more than 60 optical polarization measurements, both spectropolarimetric and broadband integrated measurements, to constrain individually all birefringent coefficients of mass dimensions $d = 4$ and $d = 5$. Constraints on the dimensionless coefficients $k_{(E)jm}^{(4)}$ and $k_{(B)jm}^{(4)}$ are $<3 \times 10^{-34}$, and the $k_{(V)jm}^{(5)}$ are constrained to $<7 \times 10^{-26}\,\mathrm{GeV}^{-1}$.

3. Summary and outlook

Astrophysical polarization measurements are an extremely powerful tool to constrain Lorentz-invariance violation in the photon sector due to the extremely long baselines.

In the future, new instruments with significantly improved high-energy polarization sensitivity will become available.[10] The IXPE satellite scheduled for launch in early 2021 will measure x-ray polarization in the 2 – 8 keV range. Compton telescopes, such as the balloon-borne COSI or the proposed AMEGO mission concept, are sensitive to polarization in the 500 keV – 5 MeV energy range. A slightly lower energy range is covered by the proposed GRB polarimeter LEAP. AdEPT is a concept for a gamma-ray pair-production telescope that would allow measuring the polarization of gamma-rays of tens of MeV. The ability to measure polarization at high energies is crucial in particular for constraining coefficients of $d \geq 6$.

References

1. D. Colladay and V.A. Kostelecký, Phys. Rev. D **55**, 6760 (1997); Phys. Rev. D **58**, 116002 (1998); V.A. Kostelecký, Phys. Rev. D **69**, 105009 (2004); V.A. Kostelecký and M. Mewes, Astrophys. J. Lett. **689**, L1 (2008).
2. V.A. Kostelecký and M. Mewes, Phys. Rev. D **80**, 015020 (2009).
3. F. Kislat and H. Krawczynski, Phys. Rev. D **95**, 083013 (2017).
4. F. Kislat, Symmetry **10**, 596 (2018).
5. F. Kislat and H. Krawczynski, Phys. Rev. D **92**, 045016 (2015).
6. E. Komatsu *et al.*, Astrophys. J. Suppl. Ser. **192**, 18 (2011); M.L. Brown *et al.*, Astrophys. J. **705**, 978 (2009); L. Pagano *et al.*, Phys. Rev. D **80**, 043522 (2009).
7. *Data Tables for Lorentz and CPT Violation*, V.A. Kostelecký and N. Russell, 2019 edition, arXiv:0801.0287v12.
8. V.A. Kostelecký and M. Mewes, Phys. Rev. Lett. **110**, 201601 (2013); A.S. Friedman *et al.*, Phys. Rev. D **99**, 035045 (2019).
9. K. Toma *et al.*, Phys. Rev. Lett. **109**, 241104 (2012); P. Laurent *et al.*, Phys. Rev. D **83**, 121301 (2011); F.W. Stecker, Astropart. Phys. **35**, 95 (2011); J.-J. Wei, Mon. Not. Royal Astron. Soc. **485**, 2401 (2019).
10. M.C. Weisskopf *et al.*, Results Phys. **6**, 1179 (2016); C.Y. Yang *et al.*, *Proc. SPIE* **10699**, 106992K (2018); A. Moiseev, PoS ICRC **2017**, 799 (2018); M. McConnell, New Astron. Rev. **76**, 1 (2017); S.D. Hunter *et al.*, Astropart. Phys. **59**, 18 (2014).

Test of Lorentz Violation with Astrophysical Neutrino Flavor at IceCube

Teppei Katori,* Carlos A. Argüelles,† Kareem Farrag,*,‡ and Shivesh Mandalia*

*Queen Mary University of London, E1 4NS, UK

†Massachusetts Institute of Technology, Cambridge, MA 02139, USA

‡University of Southampton, Southampton, SO17 1BJ, UK

On behalf of the IceCube Collaboration

Astrophysical high-energy neutrinos observed by IceCube are sensitive to small effects in a vacuum such as those motivated from quantum-gravity theories. Here, we discuss the potential sensitivity to Lorentz violation in the diffuse astrophysical neutrino data from IceCube. The estimated sensitivity reaches the Planck-scale physics motivated region providing IceCube with real discovery potential for Lorentz violation.

1. Neutrino interferometry

Neutrinos make a natural interferometric system. Their production and detection occur in their flavor eigenstates, but they propagate in their hamiltonian eigenstates. Thus, tiny disturbances during their propagation, for example tiny couplings with quantum-gravity motivated physics in the vacuum, could lead to an unexpected flavour composition at detection.[1]

Astrophysical neutrinos propagate $\sim \mathcal{O}(100)$ Mpc resulting in these neutrinos to become incoherent at detection, and so the phase information is washed out. However, an incoherent neutrino mixing is caused by an effective hamiltonian, which may include potential Lorentz-violating neutrino couplings.[2] Thus, information about Lorentz violation is imprinted on the flavour composition of astrophysical neutrinos measured at Earth. The goal of this analysis is to find nonzero SME coefficients from the flavor data of astrophysical neutrinos detected at IceCube.

2. New-physics sensitivity of astrophysical neutrino flavor

Figure 1 shows a naive estimation of the maximum sensitivity of different methods to look for Lorentz violation. Our focus is to perform the most sensitive test of Lorentz violation, hopefully reaching into the discovery region. Here, the x axis is the mass dimension d of the effective operators, and the y axis is the order of the operator scale (Λ_d) normalized to the Planck mass $(M_{\text{Planck}} \sim 1.2 \times 10^{19}\,\text{GeV})$. For example, "0" on the y axis for a dimension-six operator $(d = 6)$ corresponds to $\sim 10^{-38}\,\text{GeV}^{-2}$. Note that such definitions makes sense only for nonrenormalizable operators (operators with $d > 4$), which is traditionally used to look for new physics. A system that reaches a smaller scale in this figure has a better sensitivity to new physics. As a general trend, high-energy sources are more sensitive to higher-dimension operators. Neutrinos get extra sensitivity due to their interferometric nature. The solid line describes the naive sensitivity of astrophysical neutrino flavor physics to Lorentz violation, whose sensitivity exceeds any known sectors at $d = 5, 6, 7$. Furthermore, their sensitivities reach the expected region of Planck-scale physics for $d = 5, 6$ giving this analysis a real potential to discover quantum gravity.

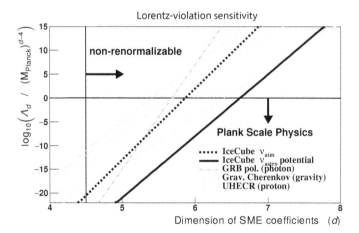

Fig. 1: Maximum-sensitivity comparison for different Lorentz-violation tests. Here, the x axis is the mass dimension of the Lorentz-violating operators, and the y axis is the new-physics scale normalized to powers of the Planck mass $(\Lambda_d/M_{Planck}^{d-4})$. The solid line is the expected sensitivity of astrophysical neutrinos in IceCube, and the dotted line is for atmospheric-neutrino limits from IceCube.[3] The dashed gray line is the limit from vacuum gravitational Cherenkov radiation,[4] the dashed-dotted gray line is from the gamma-ray-burst polarization analyses,[5] and the gray band is estimated from the ultrahigh-energy cosmic ray spectrum.[6]

3. Astrophysical neutrino-flavor triangle

The main parameter of this analysis is the fraction of observed neutrino flavors (flavor ratio) of astrophysical neutrinos ($\nu_e : \nu_\mu : \nu_\tau$) displayed in the flavor triangle ternary diagram (Fig. 2). Each corner represents a pure flavor state of the astrophysical neutrino flux; for example, the bottom right corner is pure electron composition, i.e., the ratio (1 : 0 : 0). Since we want to measure an unexpected flavor ratio due to Lorentz violation, which is represented by a point in this diagram, we must *a priori* know the flavor ratio without new physics. To predict the flavor ratio at Earth, first, we need to know the flavor ratio at production. This is expected somewhere between a ν_e dominant and a ν_μ dominant scenario (toward the right axis of the triangle). Thus, a generic model such that $(x : 1 - x : 0)$ with $0 < x < 1$ can describe all possible production models. Secondly, without assuming any new physics, neutrinos mix via their masses. We use the neutrino mass parameters from a global oscillation-data fit within the neutrino Standard Model (νSM).[7] The combination of these two makes the hatched region. Since the central region of the triangle has the highest phase-space density, any production models with reasonable assumptions (including three flavors, unitarity, etc.) will end up in the central area in this diagram.[8] We show 68% and 95% contours of the flavor-ratio measurement from an IceCube analysis.[9] Most of the hatched region is contained in this contour meaning current analyses do not have enough power to distinguish different production scenarios of astrophysical neutrinos within the νSM.

We now introduce Lorentz violation. As an example, we introduce an isotropic $d = 6$ SME coefficient with maximal ν_μ–ν_τ mixing. We assume all astrophysical neutrinos follow an $\sim E^{-2}$ spectrum between 60 TeV and 10 PeV, and the scale of the coefficient $\mathring{c}^{(6)}$ is varied from 10^{-52} GeV^{-2} (i.e., very small) to 10^{-42} GeV^{-2} (to very large). We also assume three standard astrophysical neutrino production models: ν_e dominant $(1 : 0 : 0)$, gray; ν_μ dominant $(0 : 1 : 0)$, light gray; and pion-decay models $(0.33 : 0.66 : 0)$, black. When the scale of the SME coefficient is too small, it will cause no effects on the observable flavor ratio, and all three scenarios end up in the hatched region. Once the value starts to increase, some of them start to leave the hatched region and the contour. If this is the case, we would observe a large deviation of the astrophysical neutrino flavor ratio from standard scenarios and hence could discover nonzero Lorentz violation. Clearly, experiments need to strive to shrink this contour. Improving knowledge of oscillation parameters will help to shrink the hatched region.

In this example, we expect to find nonzero SME coefficients if they cause μ–τ mixing under the assumption of a high ν_e component at the production. Note that the formalism used in this study can also be employed to look for other types of new physics.[10–13]

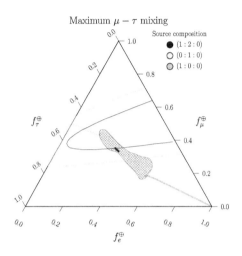

Fig. 2: Flavor triangle ternary diagram for astrophysical neutrinos. Here, the hatched region includes all possible scenarios of flavor ratios observed at Earth by assuming the production flavor ratio $(x : 1 - x : 0)$, $0 < x < 1$ and νSM oscillation parameters.[7] The 68% and 95% contours are from IceCube data.[9] Here, we show three scenarios with Lorentz violation. The black line assumes the production flavor ratio is $(0.33 : 0.66 : 0)$, whereas the gray line assumes $(1 : 0 : 0)$, and light gray line assumes $(0 : 1 : 0)$.

References

1. MiniBooNE Collaboration, T. Katori, Mod. Phys. Lett. A **27**, 1230024 (2012).
2. V.A. Kostelecký and M. Mewes, Phys. Rev. D **85**, 096005 (2012).
3. IceCube Collaboration, M.G. Aartsen *et al.*, Nature Phys. **14**, 961 (2018).
4. V.A. Kostelecký and J.D. Tasson, Phys. Lett. B **749**, 551 (2015).
5. V.A. Kostelecký and M. Mewes, Phys. Rev. Lett. **110**, no. 20, 201601 (2013).
6. L. Maccione *et al.*, J. Cosmol. Astropart. Phys. **0904**, 022 (2009).
7. I. Esteban *et al.*, J. High Energy Phys. **1901**, 106 (2019).
8. C.A. Argüelles *et al.*, Phys. Rev. Lett. **115**, 161303 (2015).
9. IceCube Collaboration, M.G. Aartsen *et al.*, Astrophys. J. **809**, 98 (2015).
10. R.W. Rasmussen *et al.*, Phys. Rev. D **96**, 083018 (2017).
11. N. Klop and S. Ando, Phys. Rev. D **97**, 063006 (2018).
12. M. Bustamante and S.K. Agarwalla, Phys. Rev. Lett. **122**, 061103 (2019).
13. Y. Farzan and S. Palomares-Ruiz, Phys. Rev. D **99**, 051702 (2019).

Pulsar Tests of Gravitational Lorentz Violation

Lijing Shao

Kavli Institute for Astronomy and Astrophysics, Peking University,
Beijing 100871, China

Pulsars are precision celestial clocks. When being put in a binary, their ticking conveys the secret of the underlying spacetime geometrodynamics. We use pulsars to test if the gravitational interaction possesses a tiny deviation from Einstein's General Relativity. In the framework of the Standard-Model Extension, we systematically search for Lorentz-violating operators cataloged by (a) the minimal couplings of mass dimension four, (b) the CPT-odd couplings of mass dimension five, and (c) couplings of mass dimension eight violating the gravitational weak equivalence principle. No deviation from General Relativity has been found yet.

1. Introduction

Pulsars are magnetized rotating neutron stars (NSs). Their stable rotations form precision celestial clocks across the sky. Though being thousands of light years away, we can use the technology of pulsar timing to obtain accurate physical inferences.[1] For a binary system, the orbital motion is determined by the gravitational interaction. Thus, a binary pulsar provides a fundamental way to look into whether Einstein's General Relativity (GR) correctly describes gravity.

In GR, spacetime is a four-dimensional differentiable manifold and the tangent space of every single point has the symmetry of local Lorentz invariance (LLI). In many searches for a fundamental quantum-gravity theory, LLI has been questioned. The so-called Standard-Model Extension (SME) presents a practically convenient way to systematically investigate the possibility of Lorentz violation.[2] Binary pulsars were proposed to be excellent laboratories to study the gravity sector of the SME.[3] In this contribution, we present a short summary of what has been achieved along this line of investigation.[4]

2. The gravity sector of the SME

In the gravity sector of the SME, generic Lorentz-violating operators are added to the lagrangian of GR. Sorted by the mass dimension of these operators, the lagrangian in the linearized limit is[2,3]

$$\mathcal{L} = \mathcal{L}_{\text{GR}} + \mathcal{L}_{\text{SME}}^{(4)} + \mathcal{L}_{\text{SME}}^{(5)} + \cdots , \tag{1}$$

where

$$\mathcal{L}_{\text{GR}} = -\frac{1}{32\pi G} h^{\mu\nu} G_{\mu\nu} + \frac{1}{2} h_{\mu\nu} T_{\text{matter}}^{\mu\nu} , \tag{2}$$

$$\mathcal{L}_{\text{SME}}^{(4)} = \frac{1}{32\pi G} \bar{s}^{\mu\kappa} h^{\nu\lambda} \mathcal{G}_{\mu\nu\kappa\lambda} , \tag{3}$$

$$\mathcal{L}_{\text{SME}}^{(5)} = -\frac{1}{128\pi G} h_{\mu\nu} q^{\mu\rho\alpha\nu\beta\sigma\gamma} \partial_\beta R_{\rho\alpha\sigma\gamma} . \tag{4}$$

In the above expressions, $\bar{s}^{\mu\kappa}$ and $q^{\mu\rho\alpha\nu\beta\sigma\gamma}$ are Lorentz-violating vacuum expectation values that arise from spontaneous symmetry breaking involving dynamical tensor fields in underlying physics.[2,3]

3. Pulsar tests

Lorentz-violating operators were constrained by precision pulsar timing; see the annually updated *Data Tables for Lorentz and CPT Violation*[5] for a comprehensive summary.

3.1. *Minimal couplings*

Lagrangian (3) gives the mass dimension four couplings in the SME. They are the lowest-order operators.[3] Because of the tensorial nature of $\bar{s}^{\mu\kappa}$, multiple binary pulsars with different locations in the sky and different orbital orientations are extremely powerful to break parameter degeneracy when constraining the various components of $\bar{s}^{\mu\kappa}$. The first study of pulsars in the gravity sector of the SME constructed 27 independent tests from 13 pulsar systems. The $\bar{s}^{\text{T}k}$ and $\bar{s}^{jk}$ ($j, k = \text{X}, \text{Y}, \text{Z}$) components are jointly constrained at the levels of $\mathcal{O}\left(10^{-9}\right)$ and $\mathcal{O}\left(10^{-11}\right)$, respectively.[4] In addition, by using the boost due to the systematic velocity between binary pulsars and the solar system, $\bar{s}^{\text{TT}}$ is constrained to be smaller than $\mathcal{O}\left(10^{-5}\right)$.[4]

3.2. *CPT-violating couplings*

Lagrangian (4) breaks CPT symmetry.[3] The modified acceleration introduced is proportional to $\sim qv \times Gm_1m_2/r^3$ where v is the relative speed between the two gravitating objects. Due to the appearance of v, static experiments in ground-based laboratories are extremely disadvantageous to place constraints on $q^{\mu\rho\alpha\nu\beta\sigma\gamma}$. In contrast, binary pulsars have a significant relative speed $v \sim 10^3\,\mathrm{km\,s^{-1}}$. In the "maximal-reach" approach, in which different components are considered to be nonvanishing on a one-by-one basis, $q^{\mu\rho\alpha\nu\beta\sigma\gamma}$ is constrained to be less than $\mathcal{O}\left(10^0\right)$–$\mathcal{O}\left(10^1\right)$ m.[4]

3.3. *Couplings violating the weak equivalence principle*

In the above subsections, only lagrangian terms quadratic in $h_{\mu\nu}$ are considered. If cubic couplings are taken into account, new phenomena appear.[3,4] The leading terms in this scenario read,

$$\frac{\sqrt{-g}}{16\pi G} k^{(8)}_{\alpha\beta\gamma\delta\kappa\lambda\mu\nu\epsilon\zeta\eta\theta} R^{\alpha\beta\gamma\delta} R^{\kappa\lambda\mu\nu} R^{\epsilon\zeta\eta\theta} \subset \mathcal{L}^{(8)}_{\mathrm{SME}}. \tag{5}$$

Such a lagrangian introduces compactness-dependent accelerations that violate the gravitational weak equivalence principle (GWEP) for self-gravitating objects. The more compact the object, the larger the acceleration corrections. NSs with compactness $\sim GM/Rc^2 \simeq 0.2$ (compared with $< 10^{-26}$ for terrestrial experiments) are excellent objects to search for GWEP-violating signals. The first empirical study of Eq. (5) was conducted with pulsar timing and set limits on $k^{(8)}_{\alpha\beta\gamma\delta\kappa\lambda\mu\nu\epsilon\zeta\eta\theta}$ at the level of $\mathcal{O}\left(10^2\right)\,\mathrm{km}^4$ in the maximal-reach approach.[4]

4. Discussion

In this contribution, pulsar tests of gravitational Lorentz violation have been reviewed concisely in the SME framework. These tests are related to the minimal couplings, the CPT-breaking coefficients as well as GWEP-violating effects. No violation has been discovered. Complementary tests searching for tiny deviations from GR have also been performed in the alternative metric-based parametrized post-Newtonian formalism.[6]

In principle, NSs are strong-field objects[1] and a strong-field version of the SME or the parametrized post-Newtonian formalism should be applied. However, generic frameworks with strong-field objects are still under development. Some strong-field phenomena were studied in specific classes of theories, e.g., scalar–tensor gravity.[7] In this sense, tests presented here are *effective* strong-field counterparts. Generally, the signals considered here within a weak-field approach are actually expected to be enhanced in strong fields. Therefore, we consider the limits reviewed here as conservative.

With more pulsars to be discovered and more precision measurements to be made by the Five-hundred-meter Aperture Spherical Telescope and the Square Kilometre Array[8] in the near future, better limits are guaranteed. If Nature exhibits deviations from Lorentz symmetry, a discovery will change our understanding of the physical world forever.

Acknowledgments

This work was supported by the Young Elite Scientists Sponsorship Program of the China Association for Science and Technology (2018QNRC001). It was partially supported by the National Natural Science Foundation of China (11721303), and the Strategic Priority Research Program of the Chinese Academy of Sciences through grant No. XDB23010200.

References

1. D.R. Lorimer and M. Kramer, *Handbook of Pulsar Astronomy*, Cambridge University Press, Cambridge, 2005; I.H. Stairs, Living Rev. Rel. **6**, 5 (2003); N. Wex, arXiv:1402.5594; L. Shao and N. Wex, Sci. China Phys. Mech. Astron. **59**, 699501 (2016); M. Kramer, Int. J. Mod. Phys. D **25**, 1630029 (2016).
2. D. Colladay and V.A. Kostelecký, Phys. Rev. D **55**, 6760 (1997); Phys. Rev. D **58**, 116002 (1998); V.A. Kostelecký, Phys. Rev. D **69**, 105009 (2004).
3. Q.G. Bailey and V.A. Kostelecký, Phys. Rev. D **74**, 045001 (2006); V.A. Kostelecký and J.D. Tasson, Phys. Rev. D **83**, 016013 (2011); Q.G. Bailey, V.A. Kostelecký, and R. Xu, Phys. Rev. D **91**, 022006 (2015); R.J. Jennings, J.D. Tasson, and S. Yang, Phys. Rev. D **92**, 125028 (2015); Q.G. Bailey, Phys. Rev. D **94**, 065029 (2016); Q.G. Bailey and D. Havert, Phys. Rev. D **96**, 064035 (2017).
4. L. Shao, Phys. Rev. Lett. **112**, 111103 (2014); Phys. Rev. D **90**, 122009 (2014); L. Shao and Q.G. Bailey, Phys. Rev. D **98**, 084049 (2018); Phys. Rev. D **99**, 084017 (2019).
5. *Data Tables for Lorentz and CPT Violation*, V.A. Kostelecký and N. Russell, 2019 edition, arXiv:0801.0287v12.
6. C.M. Will, Living Rev. Rel. **17**, 4 (2014); K. Nordtvedt, Astrophys. J. **320**, 871 (1987); T. Damour and G. Esposito-Farèse, Phys. Rev. D **46**, 4128 (1992); L. Shao and N. Wex, Class. Quant. Grav. **29**, 215018 (2012); Class. Quant. Grav. **30**, 165020 (2013); L. Shao *et al.*, Class. Quant. Grav. **30**, 165019 (2013); L. Shao, Universe **2**, 29 (2016).
7. T. Damour and G. Esposito-Farèse, Phys. Rev. Lett. **70**, 2220 (1993); Phys. Rev. D **54**, 1474 (1996); L. Shao, *et al.*, Phys. Rev. X **7**, 041025 (2017); N. Sennett, L. Shao, and J. Steinhoff, Phys. Rev. D **96**, 084019 (2017).
8. M. Kramer *et al.*, New Astron. Rev. **48**, 993 (2004); R. Nan *et al.*, Int. J. Mod. Phys. D **20**, 989 (2011); L. Shao *et al.*, Proc. Sci. AASKA14 (2015) 042; P. Bull *et al.*, arXiv:1810.02680.

CPT- and Lorentz-Violation Tests with Muon $g - 2$

B. Quinn

Department of Physics and Astronomy, University of Mississippi,
University, MS 38677, USA

On behalf of the Muon $g - 2$ Collaboration

The status of Lorentz- and CPT-violation searches using measurements of the anomalous magnetic moment of the muon is reviewed. Results from muon $g-2$ experiments have set the majority of the most stringent limits on Standard-Model Extension Lorentz and CPT violation in the muon sector. These limits are consistent with calculations of the level of Standard-Model Extension effects required to account for the current 3.7σ experiment–theory discrepancy in the muon's $g - 2$. The prospects for the new Muon $g - 2$ Experiment at Fermilab to improve upon these searches is presented.

1. The anomalous magnetic moment

The magnetic moment of the muon can be expressed by the relation

$$\vec{\mu} = g\frac{e}{2m}\vec{s} = (1 + a_\mu)\frac{e}{m}\vec{s}, \tag{1}$$

where the first term arises from the leading-order Dirac theory, and the anomaly, $a_\mu = (g - 2)/2$, represents the sum of all higher-order loop diagrams.[1] The anomalous magnetic moment includes Standard-Model (SM) terms from QED, EW, and QCD processes, as well as possible contributions from Beyond the Standard Model (BSM) physics. The muon anomaly was measured to very high precision (540 ppb) by the BNL E821 experiment yielding $a_\mu^{\text{E821}} = 116592089(63) \times 10^{-11}$.[2] When compared to the most recent SM calculations, the difference between the BNL result and theory is 3.7σ, as shown in Fig. 1.[3] This discrepancy may be a sign of new physics. The new Muon $g-2$ experiment, E989, is currently running at Fermilab and aims to measure a_μ to 140 ppb, a factor of four improvement in precision.

2. CPT- and Lorentz-violating signatures in $g - 2$

In the Muon $g - 2$ experiment, a beam of polarized muons is injected into a storage ring. The anomaly is determined by measuring the ratio

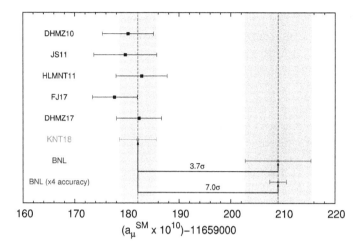

Fig. 1: Comparison of SM evaluations of a_μ with the most recent experimental result and prospect. [3]

of two frequencies: $\vec{\omega}_a = \vec{\omega}_c - \vec{\omega}_s$, which is the rate at which the muon's spin (ω_s) advances relative to its momentum (ω_c), and the proton Larmor-precession frequency ω_p, which is a measure of the ring magnetic field. [4] These frequencies are related to the anomaly by

$$a_\mu = \frac{\omega_a}{\omega_p} \frac{\mu_p}{\mu_e} \frac{m_\mu}{m_e} \frac{g_e}{2}. \tag{2}$$

The Standard-Model Extension (SME) is a general framework that describes CPT- and Lorentz-invariance violation by adding new terms to the SM lagrangian. [5] A minimal SME expression for the muon sector is

$$\mathcal{L} = -a_\kappa \bar{\psi}\gamma^\kappa \psi - b_\kappa \bar{\psi}\gamma_5 \gamma^\kappa \psi - \tfrac{1}{2} H_{\kappa\lambda} \bar{\psi}\sigma^{\kappa\lambda}\psi$$
$$+ \tfrac{1}{2} i c_{\kappa\lambda} \bar{\psi}\gamma^\kappa \overset{\leftrightarrow}{D}{}^\lambda \psi + \tfrac{1}{2} i d_{\kappa\lambda} \bar{\psi}\gamma_5 \gamma^\kappa \overset{\leftrightarrow}{D}{}^\lambda \psi. \tag{3}$$

Equation (3) predicts two Lorentz- and CPT-violating effects: a μ^+/μ^- ω_a difference, $\Delta\omega_a = \langle \omega_a^{\mu^+} \rangle - \langle \omega_a^{\mu^-} \rangle$, and a sidereal ω_a variation. [6]

In terms of the SME coefficients, $\Delta\omega_a = (4b_Z/\gamma)\cos\chi$, where χ is the colatitude of the experiment. Experimentally, it is convenient to perform the analysis on the ratio $\mathcal{R} = \omega_a/\omega_p$ from Eq. (2). The BNL E821 result $\Delta\mathcal{R} = -(3.6 \pm 3.7) \times 10^{-9}$ yields the limit $b_Z = -(1.0 \pm 1.1) \times 10^{-23}$ GeV. [7] Comparison of $\omega_a^{\mu^+}$ from one experiment with $\omega_a^{\mu^-}$ from a second at a different colatitude affords sensitivity to the d and H coefficients, and doing so with BNL E821 and an earlier muon $g-2$ experiment at CERN [8] gives $(m_\mu d_{Z0} + H_{XY}) = (1.6 \pm 5.6) \times 10^{-23}$ GeV.

Sidereal variation in ω_a is investigated using a Lomb–Scargle test[9] for a significant amplitude of ω_a oscillation at the sidereal frequency. The Lomb–Scargle method is optimized for data unequally spaced in time, as is the case for E821. The limits on such an amplitude in the BNL data are $A^{\mu^-} < 4.2\,\mathrm{ppm}$ and $A^{\mu^+} < 2.2\,\mathrm{ppm}$, which by the relationship $A^\mu = 2\check{b}^\mu_\perp \sin\chi$ is equivalent to $\check{b}^{\mu^-}_\perp \le 2.6\times10^{-24}\,\mathrm{GeV}$ and $\check{b}^{\mu^+}_\perp \le 1.4\times10^{-24}\,\mathrm{GeV}$.[7] These BNL E821 limits as well as other on both minimal ($d=4$) and nonminimal ($d\ge5$) SME coefficients[10] are the most stringent in the muon sector.

3. Prospects for E989

The goal for Muon $g-2$ at Fermilab is to reduce the BNL a_μ uncertainty by a factor of four from 540 ppb to 140 ppb. This will be achieved by utilizing Fermilab's much higher intensity muon beam to collect 21 times the BNL μ^+ statistics, and by reducing the overall systematic uncertainty by a factor of 2.5 through detector upgrades and improved analysis techniques. Reaching that goal would increase the significance of the BNL discrepancy from 3.7σ to $\sim7\sigma$ given the same central value for a_μ.

With regard to sidereal-variation Lorentz and CPT tests, sensitivity roughly scales with ω_a uncertainty. Thus, E989 should be able to reach limits of $\sim5\times10^{-25}\,\mathrm{GeV}$ and could do even better due to the possibility to search for the oscillation with a Fourier-transform method since the E989 data will be time-stamped allowing binning in equally-spaced time periods. Also, the full three-year run for Muon $g-2$ will include data for most of the calendar year. Contrary to BNL E821, which ran the same three months in each year of operation, this permits a search for annual variation in a_μ.

Obviously, to measure $\Delta\omega_a$, Muon $g-2$ needs μ^+ and μ^- data. The Fermilab Muon $g-2$ schedule features μ^+ runs extending through early 2021. The Collaboration is exploring the technical requirements to carry out a μ^- run, as was done in E821. Items to be addressed include issues related to the lower initial muon flux and the need to improve the storage-ring vacuum. The optimal time for a switchover to μ^- depends on the results of the current, approved μ^+ runs.

A μ^- run does not simply represent one more test. The μ^- data gives access to many additional SME coefficients. Furthermore, JPARC is preparing E34, a muon $g-2$ experiment with an ultra-cold muon beam.[11] E34 proposes to measure a_{μ^+} to 450 ppb. This would make possible a substantial improvement of the $(m_\mu d_{Z0} + H_{XY})$ limit for two reasons. First, the BNL/CERN 540 ppm/7000 ppm precisions would be replaced with Fermilab/JPARC 140 ppm/450 ppm. Second, this limit is proportional to $(\cos\chi_1 - \cos\chi_2)$, and there is much greater difference between Fermilab's and JPARC's colatitudes then between BNL and CERN. Because E34 utilizes muonium, and thus cannot measure a_{μ^-}, the only possibility for realizing this potential improvement is with a Fermilab Muon $g - 2$ μ^- run.

Finally, it has been shown that a nonminimal SME coefficient $\check{H}^{(5)}_{230} \simeq 3 \times 10^{-25}\,\mathrm{GeV}^{-1}$ can account for the 3.7σ discrepancy in muon $g - 2$.[12] The result from BNL E821 of $\check{H}^{(5)}_{230} = (2.9 \pm 3.0) \times 10^{-24}\,\mathrm{GeV}^{-1}$ is compatible with this level. With an E989 μ^- run and the promise of Fermilab and JPARC sensitivity goals, it may be possible to not only establish a significant discrepancy, but also make a statement concerning whether or not Lorentz and CPT violation is the physics responsible for it.

Acknowledgments

This work was supported in part by the US Department of Energy and Fermilab under contract No. DE-SC0012391. The author is grateful to Alan Kostelecký for the CPT'19 invitation and conversations.

References

1. M. Passera, J. Phys. G **31**, R75 (2005).
2. Muon $g - 2$ Collaboration, G.W. Bennett *et al.*, Phys. Rev. Lett. **92**, 161802 (2004).
3. A. Keshavarzi, D. Nomura, and T. Teubner, Phys. Rev. D **97**, 114025 (2018).
4. Muon $g - 2$ Collaboration, J. Grange *et al.*, arXiv:1501.06858.
5. D. Colladay and V.A. Kostelecký, Phys. Rev. D **58**, 116002 (1998).
6. R. Bluhm, V.A. Kostelecký, and C.D. Lane, Phys. Rev. Lett. **84**, 1098 (2000).
7. Muon $g - 2$ Collaboration, G.W. Bennett *et al.*, Phys. Rev. Lett. **100**, 091602 (2008).
8. CERN–Mainz–Daresbury Collaboration, J. Bailey *et al.*, Nucl. Phys. B **150**, 1 (1979).
9. N.R. Lomb, Astrophys. Space Sci. **39**, 447 (1976); J.D. Scargle, Astrophys. J. **263**, 835 (1982).
10. *Data Tables for Lorentz and CPT Violation*, V.A. Kostelecký and N. Russell, 2019 edition, arXiv:0801.0287v12.
11. E34 Collaboration, M. Otani, JPS Conf. Proc. **8**, 025008 (2015).
12. A.H. Gomes, A. Kostelecký, and A.J. Vargas, Phys. Rev. D **90**, 076009 (2014).

Vacuum Cherenkov Radiation for Lorentz-Violating Fermions

Marco Schreck

Departamento de Física, Universidade Federal do Maranhão,
Campus Universitário do Bacanga, São Luís - MA, 65080-805, Brazil

The current article reviews results on vacuum Cherenkov radiation obtained for modified fermions. Two classes of processes can occur that have completely distinct characteristics. The first one does not include a spin flip of the radiating fermion, whereas the second one does. A résumé will be given of the decay rates for these processes and their properties.

1. Introduction

Ordinary Cherenkov radiation in an optical medium is emitted when a charged, massive particle travels faster than the phase velocity of light in that particular medium. If the latter is the case, the polarized atoms or molecules in the vicinity of the particle trajectory emit their wave trains in phase whereupon these interfere constructively. As a result, coherent radiation is produced that can be detected far away from the source.

The modified light cone or mass-shell structure in Lorentz-violating theories may be responsible for an energy loss of a charged, massive particle in vacuum. As such a process shares certain characteristics with Cherenkov radiation in macroscopic media, it is known in the community as vacuum Cherenkov radiation. In principle, any vacuum endowed with a Lorentz-violating background field in the context of the Standard-Model Extension [1] (SME) can be interpreted as a vacuum with a nontrivial refractive index. This property explains the analogy.

Two very different scenarios exist for modified fermions [2] that may render vacuum Cherenkov radiation possible. For spin-degenerate Lorentz violation, photon emission in vacuum is possible when the slope of the mass shell at a certain energy is larger than one. Multiple photons can be emitted subsequently as long as the previously mentioned condition is satisfied. The condition fails when the particle energy drops under a certain threshold, whereupon the process ceases.

As long as the energy is sufficiently above the threshold, the momentum of the emitted photon is large enough such that the resulting recoil reverses the momentum direction of the emitting particle. A reversal of the momentum is directly connected to a helicity flip, which is why photons of helicity $h = \pm 1$ can be emitted. These photons are circularly polarized. However, when the energy of the radiating fermion is only slightly above the threshold, the photons emitted are too soft to enable a helicity change. As a consequence, they have helicity $h = 0$, i.e., they are linearly polarized.

The alternative scenario of modified fermions that allows for vacuum Cherenkov radiation is that of spin-nondegenerate Lorentz violation. In this case, photon emission is possible for a certain helicity of the fermion, i.e., the fermion loses energy and switches its mass shell. The latter corresponds to a flip of spin or helicity without a change of the momentum direction. Therefore, these particular processes are sometimes called helicity decays and photons emitted in such processes are always circularly polarized. Also, in contrast to the previous class of processes, helicity decays can occur for arbitrarily low energies of the radiating fermion.

In what follows, the decay rates[3] for both scenarios will be presented and discussed. These phenomenological results rest on a technical work[4] on the modified Dirac theory based on the nonminimal SME.

2. Spin-degenerate Lorentz violation

Let us first consider a fermion modified by spin-degenerate, isotropic Lorentz violation. The decay rates of such processes are presented in Fig. 1 as functions of the incoming fermion momentum. Minuscule values were inserted for the controlling coefficients, as those are the most interesting ones from a phenomenological perspective. Two curves are contained in the plot. The first shows the behavior of the decay rates $\Gamma_{\mathring{c},\mathring{d}}$ for the isotropic c and d coefficients, whereas the second illustrates $\Gamma_{\mathring{e},\mathring{f},\mathring{g}}$ for the isotropic e, f, and g coefficients. The thresholds for both curves are clearly visible and correspond to the momenta where the rates go to zero. For large momenta, the decay rates grow linearly with the momentum. In principle, the second curve is a scaled version of the first. The values at the axes reveal that the $\Gamma_{\mathring{e},\mathring{f},\mathring{g}}$ are suppressed by one additional power of Lorentz violation in comparison to $\Gamma_{\mathring{c},\mathring{d}}$. The thresholds are correspondingly larger. Thus, a process of this kind can occur for the dimensionless spin-nondegenerate coefficients d, g whereas it is forbidden for the dimensionful coefficients b, H.

A reasonable partial crosscheck of the results can be carried out by taking into account the coordinate transformation that maps CPT-even,

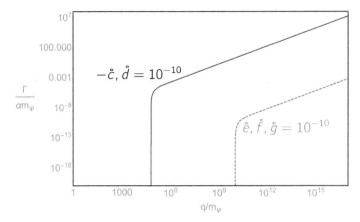

Fig. 1: Decay rates Γ for the isotropic dimensionless coefficients (indicated by rings on top of the symbols) in the SME fermion sector as a function of the fermion momentum q where m_ψ is the fermion mass and $\alpha = e^2/(4\pi)$ the fine-structure constant. The plot is double-logarithmic, whereby both axes cover a large range of values.

spin-degenerate Lorentz violation in the fermion sector to CPT-even, non-birefringent coefficients in the photon sector.[5] At leading order in the coefficients, the map[6] for the isotropic fermion coefficient $\mathring{c}$ and the isotropic photon sector coefficient $\widetilde{\kappa}_{\mathrm{tr}}$ reads $\widetilde{\kappa}_{\mathrm{tr}} = -(4/3)\mathring{c} + \ldots$. Hence, for small enough Lorentz violation, the decay rate[7] for vacuum Cherenkov radiation in the isotropic photon sector for $\widetilde{\kappa}_{\mathrm{tr}} > 0$ should correspond to the result in the isotropic fermion sector with $\mathring{c} < 0$. It was checked that the curve that relies on the isotropic photon sector perfectly matches $\Gamma_{\mathring{c}}$. This finding strengthens our confidence in the correctness of the results, as both computations were carried out completely independently from each other. Based on the fact that ultrahigh-energy cosmic rays reach Earth, constraints[3] on Lorentz violation of quarks embedded in a modified quantum electrodynamics were also obtained.

3. Spin-nondegenerate Lorentz violation

Finally, we review the behavior of a fermion modified by spin-nondegenerate Lorentz violation. In particular, the isotropic b coefficient will be discussed. The decay rate of the corresponding helicity decay is presented in Fig. 2. It is evident that the process does not have a threshold. The decay rate grows polynomially as a function of the fermion momentum until it reaches a maximum beyond which it decreases again logarithmically. Furthermore, the rate is strongly suppressed by Lorentz violation.

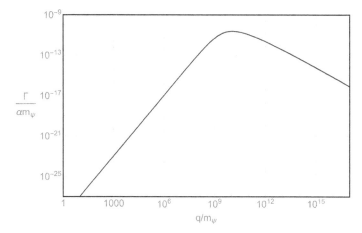

Fig. 2: Double-logarithmic plot of the decay rate for vacuum Cherenkov radiation for the isotropic b coefficient where $\mathring{b}/m_\psi = 10^{-10}$. The quantities used are the same as those in Fig. 1.

For high energies, the concept of helicity becomes more and more equivalent to chirality. Thus, the probability of a spin flip is reduced for large energies, which explains the decrease of the decay rate in the high-momentum regime. If data were available on the polarization of ultrahigh-energy cosmic rays arriving on Earth, fermion coefficients could, in fact, be constrained based on such processes.

Acknowledgments

It is a pleasure to thank the Brazilian agencies CNPq and FAPEMA for financial support under grant Nos. CNPq Universal 421566/2016-7, CNPq Produtividade 312201/2018-4, and FAPEMA Universal 01149/17. The author greatly acknowledges the hospitality of the Indiana University Center for Spacetime Symmetries (IUCSS).

References

1. D. Colladay and V.A. Kostelecký, Phys. Rev. D **58**, 116002 (1998).
2. V.A. Kostelecký and M. Mewes, Phys. Rev. D **88**, 096006 (2013).
3. M. Schreck, Phys. Rev. D **96**, 095026 (2017).
4. J.A.A.S. Reis and M. Schreck, Phys. Rev. D **95**, 075016 (2017).
5. V.A. Kostelecký and M. Mewes, Phys. Rev. D **80**, 015020 (2009).
6. B. Altschul, Phys. Rev. Lett. **98**, 041603 (2007).
7. F.R. Klinkhamer and M. Schreck, Phys. Rev. D **78**, 085026 (2008).

CPT-Violating Gravitational Orbital Perturbations

D. Colladay

New College of Florida, Sarasota, FL 34234, USA

A model for spontaneous symmetry breaking using a specific form of the bumblebee model is analyzed in the context of a Schwarzschild background metric. The resulting back reaction of the symmetry-breaking field on the metric is computed to second order. Consistency with conventional (pseudo)Riemannian geometry is demonstrated. This background field is coupled to fermions via a spin-dependent CPT-violating coupling term of a type commonly considered in the Standard-Model Extension. The perturbations of various orbital trajectories are discussed. Specifically, a spin-dependent orbital velocity is found as well as a spin-dependent precession rate.

1. Introduction

Common to most implementations of spontaneous breaking of Lorentz and CPT symmetry is the concept of some sort of potential that drives the expectation value for a vector (or tensor) field away from zero in vacuum. In a Minkowski background metric, the theory is easily understood as it produces constant background fields that can couple to various matter or force terms in the Standard Model and violate Lorentz symmetry. When the background metric is more complicated, the resulting form for the background solution is typically unknown, but can be approximated in the weak-field limit as a constant field plus small spacetime-dependent fluctuations.[1] In this talk, an example of a spontaneously generated background field in the Schwarzschild metric is considered, and the effects on the orbits of spin-coupled fermions are discussed.

2. Lagrangian for theory

The specific example used in this presentation consists of a conventional gravity term

$$\mathcal{L}_g = \frac{1}{16\pi G} eR, \tag{1}$$

a simple choice of bumblebee lagrangian[2]

$$\mathcal{L}_B = -\frac{1}{4}eB_{\mu\nu}B^{\mu\nu} - eV(B_\mu B^\mu - b^2), \quad V(x) = \lambda x, \tag{2}$$

and a spin-dependent fermion coupling term[3]

$$\mathcal{L}_f = eB_\mu\overline{\psi}\gamma^5\gamma^\mu\psi. \tag{3}$$

3. Spontaneous symmetry breaking

The bumblebee model is designed to constrain the B^μ field to be nonzero, even in the vacuum case. In the example chosen, the parameter λ in Eq. (2) is treated as a Lagrange multiplier yielding the condition $g_{\mu\nu}B^\mu B^\nu = b^2$, where the metric is chosen as Schwarzschild with $x^\mu = (t, r, \theta, \phi)$

$$d\lambda^2 = g_{\mu\nu}dx^\mu dx^\nu \tag{4}$$

$$= \left(1 - \frac{r_s}{r}\right)dt^2 - \left(1 - \frac{r_s}{r}\right)^{-1}dr^2 - r^2 d\Omega^2. \tag{5}$$

A solution for the equation of motion for B can be found by assuming local rotational invariance leading to the solution

$$B_\mu(r) = \left(b\sqrt{1 - \frac{r_s}{r}}, 0, 0, 0\right), \quad \lambda = \frac{1}{2\left(1 - \frac{r_s}{r}\right)}\left(\frac{r_s}{2r^2}\right)^2. \tag{6}$$

Note that in a local Lorentz frame, $B_a = (b, 0, 0, 0)$, where $B_a = e^\mu_a B_\mu$ are related by the vierbein field. This particular solution yields the following nonvanishing components of $B_{\mu\nu}$

$$B_{01} = -B_{10} = -\frac{r_s b}{2r^2\sqrt{1 - \frac{r_s}{r}}}, \tag{7}$$

corresponding to a radial "electric"-type field configuration.

The energy–momentum tensor for the B field can be computed using

$$T^B_{\mu\nu} = B_{\mu\alpha}B^\alpha_\nu + \frac{1}{4}B_{\alpha\beta}B^{\alpha\beta}g_{\mu\nu} - 2\lambda B_\mu B_\nu, \tag{8}$$

as

$$T^B_{\mu\nu} = -\frac{b^2}{8r^2}\left(\frac{r_s}{r}\right)^2 \tilde{g}_{\mu\nu} = -\lambda b^2 \tilde{g}_{\mu\nu}, \tag{9}$$

where $\tilde{g}_{\mu\nu}$ is the metric with the replacement $g_{11} \to -g_{11}$. Explicit calculation demonstrates that $D^\mu T^B_{\mu\nu} = 0$, which is compatible with the Bianchi identities $D_\mu G^{\mu\nu}$ of General Relativity (GR) and the relation $G^{\mu\nu} = \kappa T^{\mu\nu}$, a very desirable feature of theories involving spontaneous breaking of Lorentz invariance.[4]

184

Explicitly, the back reaction on the metric can be computed to lowest order in b^2 with the result (with $U(r) = 1 - r_s/r$)

$$g_{00} \approx U(r) \left(1 + \frac{\kappa b^2}{4U(r)} \left(\frac{r_s}{r} + (1 - \frac{r_s}{2r}) \ln U(r) \right) \right), \qquad (10)$$

and

$$g_{11} \approx -U^{-1}(r) \left(1 - \frac{\kappa b^2}{8U(r)} \frac{r_s}{r} \ln U(r) \right), \qquad (11)$$

4. Fermion coupling

Coupling to fermions as given in Eq. (3) produces an effective classical lagrangian in the form described in a previous publication:[5]

$$L_\pm^*[x, u, \tilde{e}] = -\frac{m}{2\tilde{e}} u^2 \mp \sqrt{(B \cdot u)^2 - B^2 u^2} - \frac{\tilde{e}m}{2}. \qquad (12)$$

In this expression, the metric along the worldline is generalized to $\tilde{e}$, typically referred to as an 'einbein.' The corresponding extended hamiltonian[6] is

$$\mathcal{H}_\pm^*[x, p, \tilde{e}] = -\frac{\tilde{e}}{2m} \left(p^2 - m^2 - B^2 \mp 2\sqrt{(B \cdot p)^2 - B^2 p^2} \right). \qquad (13)$$

The canonical momenta can be related to the velocities using $p_\mu = -\partial L/\partial u^\mu$ and $u^\mu = -\partial \mathcal{H}/\partial p_\mu$ yielding

$$p_0 = \frac{m}{\tilde{e}} g_{00} u^0, \qquad (14)$$

and

$$p_i = \frac{m}{\tilde{e}} g_{ij} u^j \left(1 \mp \frac{\tilde{e}b}{m\sqrt{-g_{kl}u^k u^l}} \right). \qquad (15)$$

Setting $\theta = \pi/2$ yields a hamiltonian cyclic in t and ϕ yielding conserved momenta p_t and p_ϕ. The on-shell condition $\mathcal{H}_\pm = 0$ gives the effective potential

$$V_{eff}(r) = \frac{1}{2m} \left[-\frac{r_s}{r} m^2 + \left(1 - \frac{r_s}{r}\right) \frac{p_\phi^2}{r^2} \left(1 \mp 2b \frac{(1 - \frac{r_s}{r})^{1/2}}{\sqrt{p_o^2 - (1 - \frac{r_s}{r})m^2)}} \right)^{-1} \right]. \qquad (16)$$

For vertical motion with $p_\phi = 0$, the effect of b drops out of the effective potential and the motion is equivalent to conventional free-fall motion of GR. Singular points have been previously discussed.[7] For a circular orbit

$r \to R$, a constant, setting the derivative of the potential to zero and approximating to lowest-order in b yields the orbital velocity

$$R\dot{\phi} = \sqrt{\frac{r_s}{2R(1 - \frac{3r_s}{2R})}} \pm \frac{b}{2m}\frac{(1 - \frac{r_s}{R})}{(1 - \frac{3r_s}{2R})}, \tag{17}$$

demonstrating that the geodesic motion depends on helicity. In the Newtonian limit, the difference in speeds is independent of R

$$R\dot{\phi} \approx \sqrt{\frac{GM}{R}} \pm \frac{b}{2m}, \tag{18}$$

indicating that the relative effect of b increases with the radius.

The effect on the orbital precession rate can be computed from

$$\omega_r = \frac{V''(R)}{m} = \omega_\phi \left(1 - \frac{3r_s}{R}\right)^{\frac{1}{2}} \pm \frac{b}{2mR}\frac{(1 - \frac{r_s}{R})}{\left(1 - \frac{3r_s}{R}\right)^{\frac{1}{2}}}, \tag{19}$$

the radial oscillation frequency, yielding

$$\delta_\phi \approx \frac{6\pi GM}{R} \mp \frac{\pi b}{m}\sqrt{\frac{R}{GM}}, \tag{20}$$

at leading order in r_s/R. The effect is more pronounced at larger R, but may be more difficult to observe there as well.

5. Conclusion

Local rotation invariance is useful for obtaining an example solution for a field that breaks local Lorentz invariance spontaneously in a relatively simple way. Use of a simple version of the bumblebee model produces an example that is consistent with the Bianchi identities of GR. Orbital trajectories of fermions coupled to the spontaneous-breaking field can be effected in a way that depends on the helicity of the particle, therefore breaking the equivalence principle.

References

1. V.A. Kostelecký and J.D. Tasson, Phys. Rev. D **83**, 016013 (2011).
2. A. Kostelecký and S. Samuel, Phys. Rev. Lett. **63**, 224 (1989); R. Bluhm *et al.*, Phys. Rev. D **77**, 12507 (2008).
3. D. Colladay and V.A. Kostelecký, Phys. Rev. D **55**, 6760 (1997); Phys. Rev. D **58**, 116002 (1998).
4. V.A. Kostelecký, Phys. Rev. D **69**, 105009 (2004).
5. V.A. Kostelecký and N. Russell, Phys. Lett. B **693**, 443 (2010).
6. D. Colladay, Phys. Lett. B **772**, 694 (2017).
7. D. Colladay and P. McDonald, Phys. Rev. D **92**, 085031 (2015).

Resonant Searches for Macroscopic Spin-Dependent Interactions

J.C. Long

Physics Department, Indiana University, Bloomington, IN 47405, USA

The Indiana short-range gravity experiment uses a resonant technique to investigate forces in the sub-millimeter range with femtonewton sensitivity. It has most recently been used in a search for Lorentz violation and is currently being modified to be sensitive to electron spin-dependent forces. The ARIADNE experiment, under construction at Indiana, uses an NMR technique to probe nuclear spin-dependent forces on a sample of polarized helium-3 and will have sensitivity relevant to the hypothetical axion.

1. Lorentz tests with the IU short-range gravity experiment

Short-range tests of the gravitational inverse square law have recently been used as sensitive probes of Lorentz symmetry.[1] At first glance, this application is appealing given the functional form of the corrections to the Newton force in the Standard-Model Extension (SME), a comprehensive description of Lorentz violation (LV) consistent with local field theory.[2] In addition to the usual Einstein–Hilbert term, the SME lagrangian contains an infinite series of operators of increasing mass dimension d controlling LV, leading to a set of modified field equations with a solution (in the linearized, nonrelativistic limit) given by:[3]

$$\delta U(r) = \sum_{djm} \frac{Gm_1 m_2}{r^{d-3}} Y_{jm}(\theta,\phi) k_{jm}^{N(d)}. \qquad (1)$$

Here, G is the Newton constant, m_1 and m_2 are (point) test masses, and r is the separation between them. The $Y_{jm}(\theta,\phi)$ are spherical harmonics with $j = d-2$ and $m = -j, -j+1, ..., j-1, j$. The angles θ and ϕ are the polar and azimuthal angles in the SME laboratory frame, and the $k_{jm}^{N(d)}$ are the (newtonian) coefficients of LV. For nonminimal cases ($d \geq 6$), LV in gravity leads to corrections with forces proportional to $1/r^4$ or stronger powers. These are of general relevance to short-range gravity experiments, which are typically designed with test masses of high density concentrated at small distance scales.

Results presented in Ref. 1 derive from a combined analysis of data from the short-range experiments at Huazhong University of Science and Technology (HUST) and at Indiana University (IU). The HUST experiment is described in Ref. 4. Essential features of the IU experiment, described in detail in Ref. 5, include a set of thin ($\sim 250\,\mu$m), parallel-plate, high-density (tungsten) test masses. This design concentrates as much mass as possible at the range of interest and is nominally null with respect to Newton forces. The force-generating "source" mass is driven at a resonance of the force-sensitive "detector" mass to maximize the signal, placing a heavy burden on vibration isolation and motivating the choice of a resonance near 1 kHz at which passive vibration isolation is highly effective. A stiff conducting shield between the test masses suppresses electrostatic and acoustic backgrounds but limits the average gap to about $80\,\mu$m. This design is effective for suppressing all backgrounds below the detector thermal noise, which sets the force sensitivity to about 10 fN for integration times of ~ 24 hr.[5]

The sensitivity of the results presented in Ref. 1 for $d = 6$ and $d = 8$ is lower than might be expected for simple corrections to the Newton force scaling as $1/r^4$ and $1/r^6$. This can be understood on closer examination of the force corresponding to the potential in Eq. (1). Each component of the force is described by a spherical harmonic, scaled by an LV coefficient. The point–point force between a test mass in the laboratory and another at fixed r contains equal contributions of positive and negative components as the angles θ and ϕ are varied over the half-sphere. For the planar geometry in the HUST and IU experiments, which subtends large solid angles (especially at small gaps), the LV force tends to average to zero.

Optimization for sensitivity to LV forces thus involves both minimization of test-mass separation and limiting the solid angles subtended. To make up for the corresponding overall reduction in available test mass, multiple masses can be used with periodic spacing, so as to be sensitive to only one particular phase (positive or negative) of the LV force. The HUST group has made considerable progress with this approach, with the design of a torsion balance dedicated to LV searches using test masses with alternating strips of high and low density.[6] The IU test masses could be similarly optimized, but recent efforts have been dedicated to making the experiment sensitive to spin-dependent forces.

2. The IU spin-dependent experiment

Exotic spin-dependent interactions are of interest as they could be mediated by very light bosons, which have become increasingly popular dark-matter candidates given persistent null results from WIMP searches. Two well-known examples of spin-dependent interactions include:

$$V_{MD} = g_P^1 g_S^2 \frac{\hbar^2}{8\pi m_1} \hat{\sigma}_1 \cdot \hat{r} \left(\frac{1}{\lambda r} + \frac{1}{r^2} \right) e^{-r/\lambda}, \tag{2}$$

$$V_{DD} = g_P^1 g_P^2 \frac{\hbar^3}{16\pi m_1 c^2} \left[\hat{\sigma}_1 \cdot \hat{\sigma}_2 \left(\frac{1}{\lambda r^2} + \frac{1}{r^3} \right) \right.$$
$$\left. -(\hat{\sigma}_1 \cdot \hat{r})(\hat{\sigma}_2 \cdot \hat{r}) \left(\frac{1}{\lambda r^2} + \frac{1}{r^3} \right) \right] e^{-r/\lambda}. \tag{3}$$

Here, g_S and g_P are scalar and pseudoscalar couplings, $\vec{S}_j = \hbar \hat{\sigma}_j/2$ is the spin of fermion j, $\hbar$ is Planck's constant, c is the speed of light in vacuum, $\hat{r} = \vec{r}/r$ is a unit vector along the direction between the fermions, m_1 is the mass of fermion 1, and λ is the interaction range. These are the "monopole–dipole" and "dipole–dipole" potentials investigated by Moody and Wilczek[7] as possible couplings of the axion, a light pseudoscalar motivated by the strong CP problem and a promising, light dark-matter candidate. Experimental limits allow for an unexcluded axion "window" in the distance range between $20\,\mu$m and $2\,$cm, where the IU experiment has good sensitivity. The coupling (which depends on the type of fermion) is constrained to $g_P^e g_S^N < 10^{-33}$ and $g_P^e g_P^e < 10^{-25}$, where the superscripts e and N denote electrons and nucleons, respectively.

Sensitivity to the interactions (2) and (3) requires spin-polarized masses, which are limited almost exclusively to magnetic materials. The test masses in the IU experiment have been augmented with ferrimagnets, materials with orbital compensation of the magnetization associated with the electron spins.[8] Tests at IU show that the selected material, $Dy_3Fe_5O_{12}$ (1 mm-thick pellets with 3 mm diameter suitable for test-mass attachment), exhibits total compensation of the magnetization at temperatures near 225 K.[9] Based on a molecular-field calculation, the spin density is about $4 \times 10^{20}\,\hbar/$cm^3. The corresponding strengths of the interactions (2) and (3), when compared to the 10 fN thermal noise, imply an experimental sensitivity of $g_P^e g_S^N < 10^{-20}$ and $g_P^e g_P^e < 10^{-14}$. This is about eight orders below the current bounds at $\lambda \sim 20\,\mu$m, but still well above the axion limits.

3. The ARIADNE experiment

One strategy for spin-dependent searches uses an NMR technique, in which the Larmor precession of polarized nucleons in a dilute sample is monitored for shifts as a dense, nonmagnetic source mass is brought into close proximity. The sensitivity attainable is still several orders of magnitude above indirect constraints on the axion coupling to nucleons (see, e.g., Ref. 10).

Significant improvement is possible with resonant excitation of the NMR sample at the Larmor frequency. The sample develops a transverse magnetization scaling as the T_2 relaxation time, which can be on the order of thousands of seconds. This is the principle behind the ARIADNE (Axion Resonant Interaction DetectioN Experiment) project.[11] A dense, nonmagnetic, sprocket source mass is spun for fast ($\sim 100\,\mathrm{Hz}$) modulation, and brought in close proximity to a sample of polarized helium-3 (^{3}He). The ^{3}He magnetization is monitored with a SQUID. The design is cryogenic, with the ^{3}He cell and SQUID located inside a superconducting shield. IU is a possible location for the experiment, where collaborators are building the ^{3}He system and a laboratory with complete radiofrequency shielding.

The largest of the many backgrounds to have been identified appears to be the transverse projection noise. Simulations suggest it should be possible to keep this background below $\sim 10^{-19}\,\mathrm{T}$ with a ^{3}He T_2 of $1000\,\mathrm{s}$. Assuming a $1\,\mathrm{mm}^3$ ^{3}He sample at a pressure of $1\,\mathrm{atm}$, the projected sensitivity is $g_P^N g_S^N < 10^{-33}$ at $\lambda = 1\,\mathrm{mm}$ in the region of interest to the axion.

Acknowledgments

This work is supported by US National Science Foundation grants PHY-1707986, PHY-1806757, and the IU Center for Spacetime Symmetries.

References

1. C.-G. Shao et al., Phys. Rev. Lett. **122**, 011102 (2019).
2. V.A. Kostelecký, Phys. Rev. D **69**, 105009 (2004).
3. V.A. Kostelecký and M. Mewes, Phys. Lett. B **766**, 137 (2017).
4. W.-H. Tan et al., Phys. Rev. Lett. **116**, 131101 (2016).
5. J.C. Long and V.A. Kostelecký, Phys. Rev. D **91**, 092003 (2015).
6. C.-G. Shao et al., Phys. Rev. D **94**, 104061 (2016).
7. J.E. Moody and F. Wilczek, Phys. Rev. D **30**, 130 (1984).
8. R.C. Ritter et al., Phys. Rev. D **42**, 977 (1990).
9. T.M. Leslie et al., Phys. Rev. D **89**, 114022 (2014).
10. K. Tullney et al., Phys. Rev. Lett. **111**, 100801 (2013).
11. A. Arvanitaki and A. Geraci, Phys. Rev. Lett. **113**, 161801 (2014).

The 2-Neutrino Exchange Potential with Mixing: A Probe of Neutrino Physics and CP Violation

D.E. Krause[*,†] and Q. Le Thien[*]

Physics Department, Wabash College, Crawfordsville, IN 47933, USA

†*Department of Physics and Astronomy, Purdue University, West Lafayette, IN 47907, USA*

The 2-neutrino exchange potential is a Standard-Model weak potential arising from the exchange of virtual neutrino–antineutrino pairs which must include all neutrino properties, including the number of flavors, their masses, fermionic nature (Dirac or Majorana), and CP violation. We describe a new approach for calculating the spin-independent 2-neutrino exchange potential, including the mixing of three neutrino mass states and CP violation.

The neutrino sector of the Standard Model (SM) holds great potential for revealing new physics. Interestingly, the unsolved problems of neutrino physics [e.g., the masses of the three neutrino mass states, the neutrino's fermionic nature (Dirac or Majorana), number of flavors, existence of sterile neutrinos and CP violation] all impact the 2-neutrino exchange potential (2-NEP), the weak interaction force arising from the exchange of virtual neutrino–antineutrino pairs. The formulas for the single neutrino flavor 2-NEP were first derived by Feinberg and Sucher[1] and Fischbach[2] assuming massless and massive neutrinos, respectively. A number of other authors have also investigated the 2-NEP.[3–5] Lusignoli and Petrarca[6] developed an integral formula for the 2-NEP with mixing of three neutrino flavors, but did not include all of the electroweak contributions. Here, we describe a new derivation of the 2-NEP with mixing that incorporates neutral-current (NC) and charged-current (CC) weak interactions and CP violation, and discuss the possibilities of using the 2-NEP as a probe of neutrino physics.[7]

In our approach for calculating the 2-NEP, we express the neutrino fields in the Schrödinger picture and then use time-independent perturbation theory to calculate the second-order energy shift of the neutrino-field vacuum energy due to the presence of two stationary fermions. We ignore infinite self-energy corrections, which only depend on the positions of a single fermion. The spin-independent contribution, which depends on the separation distance r, is finite, and for the single-flavor case involving NC weak interactions is found to be given by[7]

$$V_{\nu,\bar{\nu}}(r) = \frac{G_F^2 g_{V,1}^f g_{V,2}^f m_\nu^3}{4\pi^3 r^2} K_3(2m_\nu r), \qquad (1)$$

where G_F is the Fermi constant, g_V is the vector coupling constant, m_ν is the neutrino mass, and $K_n(x)$ is the modified Bessel function.

To incorporate mixing, we write the three flavor fields $\nu_\alpha(\vec{r})$, ($\alpha = e, \mu, \tau$) as linear combinations of the three mass fields $\nu_a(\vec{r})$, ($a = 1, 2, 3$), i.e., $\nu_\alpha(\vec{r}) \equiv \sum_{a=1}^3 U_{\alpha a} \nu_a(\vec{r})$, where $U_{\alpha a}$ are components of the Pontecorvo–Maki–Nakagawa–Sakata (PMNS) matrix,

$$U_{\alpha a} = \begin{pmatrix} c_{12}c_{13} & s_{12}c_{13} & s_{13}e^{-i\delta_{\mathrm{CP}}} \\ -s_{12}c_{23} - c_{12}s_{23}s_{13}e^{i\delta_{\mathrm{CP}}} & c_{12}c_{23} - s_{12}s_{23}s_{13}e^{i\delta_{\mathrm{CP}}} & s_{23}c_{13} \\ s_{12}s_{23} - c_{12}c_{23}s_{13}e^{i\delta_{\mathrm{CP}}} & -c_{12}s_{23} - s_{12}c_{23}s_{13}e^{i\delta_{\mathrm{CP}}} & c_{23}c_{13} \end{pmatrix}. \qquad (2)$$

Here, $s_{ab} = \sin\theta_{ab}$, $c_{ab} = \cos\theta_{ab}$, and δ_{CP} is the CP-violation phase. For the purpose of our calculation, we note that nucleons interact with the neutrino only via NC interactions, while leptons also require the inclusion of CC interactions. We will therefore need to consider three cases for the interaction potentials: nucleon–nucleon, nucleon–lepton, and lepton–lepton.

The 2-NEP between nucleons is the simplest since the NC current interaction is independent of neutrino flavor. For nucleons #1 and #2, we merely sum Eq. (1) over the three mass states, which gives[7]

$$V_{N_1,N_2}(r) = \frac{G_F^2 g_{V,1}^{N_1} g_{V,2}^{N_2}}{4\pi^3 r^2} \sum_{a=1}^3 m_a^3 K_3(2m_a r), \qquad (3)$$

where N = proton or neutron, $g_V^N = \frac{1}{2} - 2\sin^2\theta_W$ for protons and $g_V^N = -\frac{1}{2}$ for neutrons, where θ_W is the Weinberg angle. The magnitude of this interaction is quite small. For two neutrons, the gravitational force is larger than the 2-NEP force when $r \gtrsim 1\,\mathrm{nm}$.

For the case of a nucleon interacting with a lepton, one finds a result similar to the nucleon–nucleon potential except for a change of the lepton vector coupling, which depends on the PMNS matrix element corresponding to the lepton flavor,

$$V_{N\alpha}(r) = \frac{G_F^2 g_V^N}{4\pi^3 r^2} \sum_{a=1}^3 m_a^3 \left(g_V^\alpha + |U_{\alpha a}|^2 \right) K_3(2m_a r). \qquad (4)$$

The final case, the lepton–lepton 2-NEP, is the most interesting since the mixing has the greatest impact on the form of the potential and it involves both NC and CC interactions, but we were unable to obtain a closed-form expression if all the masses are nonzero. Instead, for two lepton flavors α and β, we found an expansion for the 2-NEP in powers of (m_-^{ab}/m_+^{ab}), where $m_\pm^{ab} \equiv m_a \pm m_b$:[7]

$$V_{\alpha\beta}(r) = \frac{G_F^2}{4\pi^3 r^2} \sum_{a=1}^{3} \left[m_a^3 \left(g_V^\alpha + |U_{\alpha a}|^2 \right) \left(g_V^\beta + |U_{\beta a}|^2 \right) K_3(2m_a r) \right] + V_{\alpha\beta,\mathrm{mix}}(r),$$ (5)

where to order $\left[\left(\frac{m_-^{ab}}{m_+^{ab}} \right)^2 \right]$

$$V_{\alpha\beta,\mathrm{mix}}(r) \simeq \frac{G_F^2}{4\pi^3 r^2} \left[\sum_{a>b}^{3} \frac{\mathrm{Re}\left(U_{\alpha a}^* U_{\alpha b} U_{\beta b}^* U_{\beta a} \right)}{4} \right.$$

$$\left. \left\{ m_+^{ab} \left[\left(m_+^{ab} \right)^2 + \left(m_-^{ab} \right)^2 \right] K_3 \left(m_+^{ab} r \right) - \frac{4\left(m_-^{ab} \right)^2}{r} K_2 \left(m_+^{ab} r \right) \right\} \right].$$ (6)

The lepton–lepton 2-NEP is particularly interesting because of its dependence on the CP-violating phase δ_{CP}. Presently, there is growing evidence[8] that $\delta_{\mathrm{CP}} \neq 0$. We show that one can write[7]

$$V_{\alpha\beta}(r) = V_{\alpha\beta}^{(0)}(r) + V_{\alpha\beta}^{(\mathrm{CP})}(r) \sin^2 \left(\frac{\delta_{\mathrm{CP}}}{2} \right),$$ (7)

where $V_{\alpha\beta}^{(0)}$ and $V_{\alpha\beta}^{(\mathrm{CP})}(r)$ are complicated functions of r independent of δ_{CP}, and $V_{ee}^{(\mathrm{CP})}(r) = 0$ by the definition of δ_{CP}. Therefore, except for the interaction between two electrons, the 2-NEP potential will depend on the CP-violating phase due to the interference of the PMNS matrix elements.

While the original study of 2-NEP arose mainly out of theoretical interests, our results raise the possibility of opening new avenues for experimental explorations of basic neutrino parameters. Because the 2-NEP involves the exchange of virtual neutrinos, all neutrino properties and energies must contribute. Besides being sensitive to the mixing angles, the 2-NEP depends directly on the actual neutrino masses, not the difference in mass squared as in neutrino-oscillation experiments. If we assume the neutrino mass spectrum lies in the range $1\,\mathrm{meV} \lesssim m_\nu \lesssim 1\,\mathrm{eV}$, the most promising experiments need to focus on the corresponding separations $1\,\mathrm{nm} \lesssim r \lesssim 1\,\mu\mathrm{m}$. The dependence of the 2-NEP on CP violation is also novel, the only long-range SM force with this property. Muonium is the natural system to explore these effects since the 2-NEP CP violation requires interactions between leptons other than between just electrons. In addition, as pointed out by Fischbach *et al.*, the contribution of the 2-NEP to the nuclear binding energy provides an interesting test of the WEP as applied to neutrinos.[9] Of course, the experimental challenges in realizing these possibilities are significant, requiring measuring forces of less than gravitational strength on the nanometer scale or the Weak Equivalence Principle at the level of $\sim 10^{-17}$. Recently, Stadnik[5] examined the potential of using high-precision spectroscopy, although some of the assumptions made may be overly optimistic.[10]

This work only touches on some of the interesting questions raised by the 2-NEP. We assumed neutrinos were Dirac fermions, examined only the spin-independent interaction, and assumed the simplest neutrino vacuum state. Alternatives to these assumptions and others involving theories beyond the SM will certainly impact the 2-NEP, providing new directions for theoretical and experimental exploration.

Acknowledgments

We thank Ephraim Fischbach for useful conversations and earlier papers on the 2-NEP, which provided significant motivation for our work. We also thank Sheakha Aldaihan and Mike Snow for discussions on the derivation of potentials, which influenced our approach.

References

1. G. Feinberg and J. Sucher, Phys. Rev. **166**, 1638 (1968); G. Feinberg, J. Sucher, and C.-K. Au, Phys. Rep. **180**, 83 (1989).
2. E. Fischbach, Ann. Phys. (NY) **247**, 213 (1996).
3. J.B. Hartle, Phys. Rev. D **1**, 394 (1970); Phys. Rev. D **49**, 4951 (1994); A. Segarra, arXiv:1606.05087; IOP Conf. Series: Journal of Physics: Conf. Series **888**, 012199 (2017); arXiv:1712.01049.
4. J.A. Grifols, E. Massó, and R. Toldrá, Phys. Lett. B **389**, 563 (1996).
5. Y.V. Stadnik, Phys. Rev. Lett. **120**, 223202 (2018).
6. M. Lusignoli and S. Petrarca, Eur. Phys. J. C **71**, 1568 (2011).
7. Q. Le Thien and D.E. Krause, Phys. Rev. D (in press); arXiv:1901.05345.
8. K. Abe, *et al.*, Phys. Rev. Lett. **121**, 171802 (2018).
9. E. Fischbach, D.E. Krause, C. Talmadge, and D. Tadić, Phys. Rev. D **52**, 5417 (1995).
10. T. Asaka, M. Tanaka, K. Tsumura, and M. Yoshimua, arXiv:1810.05429.

The Leading Trilinear Photon Coupling in the SME

Hannah J. Day[*] and Ralf Lehnert[†,‡]

*Homer L. Dodge Department of Physics and Astronomy
University of Oklahoma, Norman, OK 73019, USA*

†*Indiana University Center for Spacetime Symmetries, Bloomington, IN 47405, USA*

‡*Leibniz Universität Hannover, Welfengarten 1, 30167 Hannover, Germany*

The full Standard-Model Extension contains operators of arbitrary mass dimension. The recent classification of these operators for gauge theories has substantially broadened the avenue for phenomenological explorations of Lorentz symmetry in electrodynamics. This work focuses on the cubic mass-dimension six contribution to pure Maxwell theory. Its effects on the propagation of light in an external magnetic field are discussed, and the prospects of determining experimental constraints are assessed.

1. Introduction

The past two decades have witnessed rapid progress in experimental investigations of Lorentz and CPT symmetry.[1] The analysis and interpretation of these results relies on the Lorentz- and CPT-violating Standard-Model Extension (SME),[2,3] a framework based on effective field theory. It is therefore natural to classify the various types of Lorentz and CPT breaking according to the mass dimension d of the corresponding operator in the SME Lagrange density.

The power-counting renormalizable terms $d \leq 4$ have formed the foundation of the original minimal SME. The majority of the $d > 4$ operators governing free propagation, as well as some $4 < d \leq 6$ contributions to particle interactions, have meanwhile also been classified and exposed to experimental scrutiny.[4] However, the general SME corrections to gauge-field theories have only recently been established, expanding the range of Lorentz-violating signals and providing an impetus for intensified phenomenological and experimental SME analyses.[5]

Motivated by these developments, we focus in this work on the leading nonlinear SME contribution to pure electrodynamics. We derive the induced modifications to the Maxwell equations and employ them to investigate the corresponding SME effects on the propagation of electromagnetic waves in the presence of an external homogeneous magnetic field. We show that these effects generally include birefringence measurable in laboratory experiments. This experimental signature can serve as a basis for placing limits on this type of Lorentz breakdown with existing and future experimental efforts.

2. Model

We consider the SME limit of the usual Maxwell Lagrange density augmented by the mass-dimension six Lorentz-violating cubic interaction term given by[5]

$$\delta\mathcal{L} = -\tfrac{1}{12}(k_F^{(6)})^{\kappa\lambda\mu\nu\rho\sigma} F_{\kappa\lambda} F_{\mu\nu} F_{\rho\sigma} \,. \tag{1}$$

Here, $F_{\mu\nu} = \partial_\mu A_\nu - \partial_\nu A_\mu$ represents the ordinary electromagnetic field-strength tensor with A_μ being the conventional four-potential.

The SME coefficient $(k_F^{(6)})^{\kappa\lambda\mu\nu\rho\sigma}$ parametrizes the departure from Lorentz symmetry while maintaining CPT invariance. We are primarily interested in the flat-spacetime limit, where we may take $k_F^{(6)}$ as nondynamical and constant. This is a standard assumption and maintains the conservation of four-momentum. The structure of Eq. (1) implies that $(k_F^{(6)})^{\kappa\lambda\mu\nu\rho\sigma}$ must exhibit various index symmetries that reduce the number of independent $k_F^{(6)}$ components to 56; it can be shown that none of these represent Lorentz-invariant pieces. A possible physical origin of six of these components is noncommutative field theory.[6]

With the model lagrangian at hand, the inhomogeneous Maxwell equations are modified as follows:

$$\partial_\kappa \left[F^{\kappa\lambda} + \tfrac{1}{2}(k_F^{(6)})^{\kappa\lambda\mu\nu\rho\sigma} F_{\mu\nu} F_{\rho\sigma} \right] = j^\lambda \,, \tag{2}$$

where j^λ denotes a conserved, externally prescribed current density. The homogeneous Maxwell equations remain unaffected because they reflect the conventional relationship between field strength and potential. Equation (2) governs the main effects of this type of Lorentz violation and therefore provides a suitable basis for many phenomenological studies of $k_F^{(6)}$.

3. Light propagation in an external magnetic field

In the following, we focus on physical situations in which the total electromagnetic field can be decomposed into a given background $F_0^{\mu\nu}$ and small, localized disturbances $F^{\mu\nu}$ that can be treated dynamically as perturbations to $F_0^{\mu\nu}$. Keeping only leading-order terms in the perturbation $F^{\mu\nu}$, Eq. (2) then takes form

$$\partial_\mu \left[F^{\mu\nu} + (k_{\text{eff}})^{\mu\nu\rho\sigma} F_{\rho\sigma} \right] = 0 \tag{3}$$

in source-free regions. In this equation, we have introduced the effective coefficient

$$(k_{\text{eff}})_{\kappa\lambda\mu\nu} := (k_F^{(6)})_{\kappa\lambda\mu\nu\rho\sigma} F_0^{\rho\sigma} . \tag{4}$$

Note that Eq. (3) is linear, allowing plane-wave solutions if $F_0^{\mu\nu}$ varies slowly enough. Note also that the effective Lorentz-violating coefficient scales linearly with $F_0^{\mu\nu}$, so that $k_F^{(6)}$ effects will typically become enhanced in larger background electromagnetic fields.

For $F_0^{\mu\nu} = $ const., the equation of motion (3) is identical in structure to that generated by the SME's $(k_F)^{\mu\nu\rho\sigma}$ photon coefficient. Its plane-wave solutions $F^{\mu\nu}(x) = F^{\mu\nu}(p)e^{-ip_\beta x^\beta}$ have already been studied in detail,[7] and we may employ these results for our present purposes. In particular, the dispersion relation for the plane-wave momentum $p^\beta = (p^0, \vec{p})$ is

$$p_\pm^0 = (1 + \rho \pm \sigma)|\vec{p}| , \tag{5}$$

where $\rho = -\frac{1}{2}\hat{k}_\beta{}^\beta$ and $\sigma^2 = \frac{1}{2}(\hat{k}_{\mu\nu})^2 - \rho^2$, with $\hat{k}^{\kappa\lambda} \equiv (k_{\text{eff}})^{\kappa\mu\lambda\nu}\hat{p}_\mu\hat{p}_\nu$ and $\hat{p}^\mu \equiv (1, \vec{p}/|\vec{p}|)$. This dispersion relation is valid at leading order in $k_F^{(6)}$; it describes two independently propagating degrees of freedom that exhibit small perturbations relative to the usual two modes. The physical characteristics of such waves is in general both direction and boost dependent, as expected in the presence of Lorentz violation. Note that for $\sigma \neq 0$ the two modes are nondegenerate, so that propagation becomes birefringent.

4. Experimental prospects

A nonlinear Lorentz-violating modification of electrodynamics, such as that in Eq. (1), is expected to be amenable to tests in a broad range of physical systems. In this section, we focus on the possibility of birefringent propagation in a homogeneous external magnetic field $\vec{B}_0$, as deduced above. Existing and future experimental efforts involving laser light propagating inside an applied $\vec{B}$ field have originally been developed for measurements of vacuum magnetic birefringence.[8,9] This is a conventional, extremely small quantum-electrodynamics effect arising from nonlinear radiative corrections to the photon sector. Such an experimental system would also be affected by our $k_F^{(6)}$ birefringence and should therefore be well-suited to search for this Lorentz-violating effect.

To estimate the reach of such experiments for $k_F^{(6)}$ birefringence, we may characterize the size of this effect by the difference in the effective refractive indices $\Delta n \equiv n_- - n_+$ between the two propagating modes, the usual figure of merit for such measurements. Equation (5) implies

$$\Delta n = |\vec{B}_0| \sqrt{2\big[(k_F^{(6)})_{\alpha\mu\beta\nu\rho\sigma} \hat{f}_0^{\rho\sigma} \hat{p}^\mu \hat{p}^\nu\big]^2 - \big[(k_F^{(6)})^\alpha{}_{\mu\alpha\nu\rho\sigma} \hat{f}_0^{\rho\sigma} \hat{p}^\mu \hat{p}^\nu\big]^2}, \quad (6)$$

where $\hat{f}_0^{\mu\nu} \equiv F_0^{\mu\nu}/|\vec{B}_0|$ denotes the $\vec{B}_0$ field-strength tensor in units of $|\vec{B}_0|$. For a measurement with $\Delta n \lesssim 4.5 \times 10^{-22}$ in a $|\vec{B}_0| \simeq 2.5\,\text{T}$ magnetic field, such as PVLAS,[8] a reach of

$$|k_F^{(6)}| \lesssim 10^{-7}\,\text{GeV}^{-2} \quad (7)$$

on certain components of $k_F^{(6)}$ would seem to be achievable with a dedicated analysis of experimental data. This would represent the first observational constraint on this SME coefficient.

Acknowledgments

This work was supported in part by the US National Science Foundation under the REU program and by the Indiana University Center for Spacetime Symmetries under an IUCRG grant.

References

1. V.A. Kostelecký and N. Russell, arXiv:0801.0287v12.
2. D. Colladay and V.A. Kostelecký, Phys. Rev. D **55**, 6760 (1997); Phys. Rev. D **58**, 116002 (1998).
3. V.A. Kostelecký, Phys. Rev. D **69**, 105009 (2004).
4. V.A. Kostelecký and M. Mewes, Phys. Rev. D **80**, 015020 (2009); Phys. Rev. D **88**, 096006 (2013); Phys. Rev. D **85**, 096005 (2012); Phys. Lett. B **779**, 136 (2018); Y. Ding and V.A. Kostelecký, Phys. Rev. D **94**, 056008 (2016).
5. V.A. Kostelecký and Z. Li, Phys. Rev. D **99**, 056016 (2019); Z. Li, these proceedings.
6. S.M. Carroll *et al.*, Phys. Rev. Lett. **87**, 141601 (2001).
7. V.A. Kostelecký and M. Mewes, Phys. Rev. Lett. **87**, 251304 (2001).
8. F. Della Valle *et al.*, Eur. Phys. J. C **76**, 24 (2016).
9. P. Pugnat *et al.*, OSQAR Proposal 2006, CERN-SPSC-2006-035.

A 3+1 Decomposition of the Gravitational Sector
of the Minimal Standard-Model Extension

Nils A. Nilsson,* Kellie O'Neal-Ault,[†] and Quentin G. Bailey[†]

*National Centre for Nuclear Research, Pasteura 7, 05-077, Warsaw, Poland

[†]Embry-Riddle Aeronautical University,
3700 Willow Creek Road, Prescott, AZ 86301, USA

The 3+1 (ADM) formulation of General Relativity is used, for example, in canonical quantum gravity and numerical relativity. Here, we present a 3+1 decomposition of the minimal Standard-Model Extension gravity lagrangian. By choosing the leaves of foliation to lie along a timelike vector field we write the theory in a form that will allow for comparison and matching to other gravity models.

1. Introduction

Local Lorentz invariance is one of the cornerstones of General Relativity (GR) and modern physics. As such, it is an excellent probe of new physics, and Lorentz violation is a large and active area of research.[1] The Standard-Model Extension (SME) is an often-used effective-field-theory framework that includes all Lorentz- and CPT-violating terms.[2-4]

The 3+1 (ADM) version of GR is used, for example, in canonical quantum gravity and numerical relativity.[5,6] Here, we present a 3+1 decomposition of the minimal SME gravity lagrangian in the case of explicit Lorentz-symmetry breaking. By choosing the hypersurfaces to be spatial, we write the framework in a form that will allow for comparison and matching to other gravity models.

2. The decomposition

Using the ADM variables, the metric reads:

$$ds^2 = -N^2 dt^2 + \gamma_{ij} \left(dx^i + N^i dt \right) \left(dx^j + N^j dt \right), \qquad (1)$$

where N is the lapse function and N^i is the shift vector. These ADM variables relate points on different constant-time hypersurfaces (see Fig. 1).

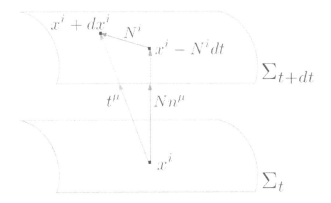

Fig. 1: Constant-time hypersurfaces Σ along with the ADM variables.

Decomposition of the manifold $\mathcal{M} \to \Sigma \times \mathbb{R}$ induces the metric $\gamma^{\mu\nu} = g^{\mu\nu} + n^\mu n^\nu$, where $n^\mu = (1/N, -N^i/N)$ is a vector normal to the foliation. The minimal gravitational sector of the SME reads as:[4,7]

$$\mathcal{L}_{\text{mSME}} = \frac{\sqrt{-g}}{2\kappa} \left[-uR + s^{\mu\nu} R_{\mu\nu}^T + t^{\mu\nu\alpha\beta} W_{\mu\nu\alpha\beta} \right], \qquad (2)$$

where $\kappa = 8\pi G$, $R_{\mu\nu}^T$ is the trace-free Ricci tensor, and $W_{\mu\nu\alpha\beta}$ is the Weyl tensor. In the isotropic limit, we can write the above lagrangian as:

$$\mathcal{L}_{\text{mSME,iso}} = \frac{\sqrt{-g}}{2\kappa} \left[{}^{(4)}s_{\mu\nu}\, {}^{(4)}R^{\mu\nu} \right], \qquad (3)$$

where a superscript (4) denotes quantities defined on $\mathcal{M}$. Here, we focus on explicit symmetry breaking, so that dynamical terms in the action vanish.[8] The above lagrangian can be rewritten as:

$$\mathcal{L}_{\text{mSME,iso}} = \frac{\sqrt{-g}}{2\kappa} {}^{(4)}s^{\mu\nu} \left[\gamma^\alpha{}_\mu \gamma^\beta{}_\nu\, {}^{(4)}R_{\alpha\beta} + n_\mu n_\nu n^\alpha n^\beta\, {}^{(4)}R_{\alpha\beta} \right.$$
$$\left. - 2\gamma^\alpha{}_\mu n_\nu n^\beta\, {}^{(4)}R_{\alpha\beta} \right], \qquad (4)$$

and by using the Gauss, Gauss–Codazzi, and Ricci equations, we can write down the fully decomposed formulation of the gravitational sector (GR + minimal SME):

$$\mathcal{L}_{\text{EH}} + \mathcal{L}_{\text{mSME,iso}} = \frac{\sqrt{-g}}{2\kappa} \left[\left(1 + n^\alpha n^\beta\, {}^{(4)}s_{\alpha\beta} \right) \left(K^{\alpha\beta} K_{\alpha\beta} - K^2 \right) \right.$$
$$\left. + \left(\tfrac{1}{3}\, {}^{(4)}s^i{}_i \right) \mathcal{R} + \nabla_\mu \left(1 + \tfrac{1}{3}\, {}^{(4)}s^i{}_i - n^\alpha n^\beta\, {}^{(4)}s_{\alpha\beta} \right) \left(a^\mu + n^\mu K \right) \right], \qquad (5)$$

where $^{(4)}s_i^i$ is the trace of the spatial part of $s^{\mu\nu}$ and $K_{\mu\nu}$ denotes the extrinsic curvature of the foliation. Moreover, we define the *acceleration vector* $a_\mu = D_\mu \ln N$ and the three-dimensional Ricci scalar $\mathcal{R}$. GR is recovered when $s^{\mu\nu} \to 0$.

3. Discussion and conclusions

Using standard tools in numerical relativity theory we have derived a 3+1 decomposition of the minimal SME gravity lagrangian in the isotropic limit. We make no linearized-gravity approximations, and thus this is an *exact* result. This complements other exact studies of the SME.[9] Our results can be used in ongoing work on identifying the dynamical degrees of freedom in the explicit-symmetry-breaking case and on matching to proposed models of quantum gravity.

Acknowledgments

NAN was partly supported by NCBJ Young Scientist Grant MNiSW 212737/E-78/M/2018. QGB and KO acknowledge support from the US National Science Foundation under grant No. 1806871 and support of Embry-Riddle Aeronautical University.

References

1. *Data Tables for Lorentz and CPT Violation,* V.A. Kostelecký and N. Russell, 2019 edition, arXiv:0801.0287v12.
2. V.A. Kostelecký and D. Colladay, Phys. Rev. D **55**, 6760 (1997).
3. V.A. Kostelecký and D. Colladay, Phys. Rev. D **58**, 116002 (1998).
4. V.A. Kostelecký, Phys. Rev. D **69**, 105009 (2004).
5. T.W. Baumgarte and S.L. Shapiro, *Numerical Relativity: Solving Einstein's Equations on the Computer*, Cambridge University Press, Cambridge, 2010.
6. C. Kiefer, *Quantum Gravity*, Oxford University Press, Oxford, 2012.
7. Q.G. Bailey and V.A. Kostelecký, Phys. Rev. D **74**, 045001 (2006).
8. R. Bluhm, Symmetry **9**, 230 (2017).
9. Y. Bonder, Phys. Rev. D **91**, 125002 (2015).

Progress Towards Ramsey Hyperfine Spectroscopy in ASACUSA

A. Nanda

Stefan Meyer Institute, Austrian Academy of Sciences,
Boltzmanngasse 3, 1090 Vienna, Austria

On behalf of the ASACUSA-CUSP Collaboration[*]

This proceedings contribution reports on progress on the design of a Ramsey-type spectrometer that will be used for the spectroscopy of the ground-state hyperfine structure of both hydrogen and deuterium.

1. Antihydrogen spectroscopy program of ASACUSA

The ASACUSA-CUSP (Atomic Spectroscopy And Collisions Using Slow Antiprotons) antihydrogen experiment based at the Antiproton Decelerator facility of CERN aims to perform precise tests of CPT symmetry by measuring the ground-state hyperfine structure (GS-HFS) of antihydrogen atoms and comparing it with that of hydrogen.[1] A brief account on the status of ASACUSA's antihydrogen program can be found within these proceedings.[2] The currently applied Rabi-type beam-spectroscopy method can be made more precise by the implementation of Ramsey's technique of using two separated oscillatory fields.[3] This improvement will first be tested on hydrogen and later on deuterium, which can help constrain the associated SME coefficients.[4]

2. Design of the spectrometer and its scope

An in-beam spectroscopy method comprises a spin-selected atomic beam, an oscillating field to drive transitions between two different spin states, a field gradient to select spin states of interest, and a detection scheme to measure the amount of beam in this state.

In this section we report the progress on a new broadband microwave-generation system that is well suited for in-beam spectroscopy using the

---

[*]http://asacusa.web.cern.ch/ASACUSA/asacusaweb/antihydrogen_cusp/main.shtml

Ramsey method. Simulations for a stripline traveling-wave device have been studied with COMSOL 5.3 using the 2D electrostatic study feature in the AC/DC Module.[5] A cross-sectional view of the device consisting of two electrodes and housed inside a 100 mm diameter cylinder is shown in Fig. 1. The 40 mm diameter circle in between the two electrodes repre-

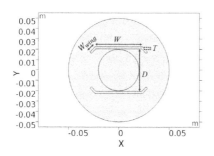

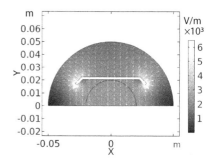

Fig. 1: Cross-sectional view of the strip-line device.

Fig. 2: Electric field in the simulated structure.

sents the area where the hydrogen beam will be present. The electrostatic problem was solved for only half the structure and using the central plane between the two electrodes as a virtual ground. The potential at the electrode boundaries was 50 V while the remaining boundaries were all defined as ground. The resulting electric field is shown in Fig. 2. The electric field is perpendicular to the electrode, hence when operated in transverse electromagnetic (TEM) mode, the oscillating magnetic field (B_{osc}) will be parallel to the electrode and perpendicular to the axis of the cylindrical housing. The impedance of the structure was calculated as $Z = 1/(c \cdot K)$, where c is the speed of light in free space and K is the computed capacitance. The electric-field inhomogeneity (ratio of the standard deviation to the average of the norm of the electric field) of the structure was studied in the area where the hydrogen beam would be present. The geometrical configuration with $W = 47$ mm, $D = 41$ mm, $T = 2$ mm, and $W_{wing} = 7$ mm has an impedance of $50.037\,\Omega$, and 0.67% electric-field inhomogeneity.

The σ and π transitions[2] require different orientations of B_{osc} relative to an external static magnetic field (B_{ext}).[2] This means that with the stripline structure and an axial magnetic field along the beam direction, we can only measure the π transition. To measure the σ transition, a TE_{110} mode cavity can be used, which has B_{osc} along its axis. As the σ transition in hydrogen is insensitive to Lorentz violation,[4] it can be used as a benchmark against the π transitions.

Due to the broadband nature of the TEM travelling waveguide we can measure the GS-HFS of deuterium around 330 MHz, in addition to that of hydrogen at 1.42 GHz. The energy shifts of the hyperfine Zeeman levels in deuterium due to the effects of CPT and Lorentz violation within the Standard-Model Extension framework are shown in Eq. (123) of Ref. 4, and they depend on the expectation values the momentum of the proton relative to the center of mass of the deuteron ($\mathbf{p}_{pd}$) and its higher powers ($\langle \mathbf{p}_{pd}^k \rangle$).[4] This feature, being significantly different from that of hydrogen, leads to the enhancement of the sensitivity to coefficients for Lorentz and CPT violation[4] by a factor of 10^9 for coefficients with $k = 2$ and 10^{18} for coefficients with $k = 4$.

3. Summary and outlook

We have reported on the geometry of a stripline structure that can be used as an interaction zone in Ramsey-type spectroscopy for GS-HFS measurements of hydrogen and deuterium. The advantages of using two different microwave regions for the σ and π transitions have also been discussed.

Next, the transition from RF feedthroughs to the striplines for feeding microwaves into the device will be designed with input from further simulations. Later, possibilities for measuring σ transitions in deuterium will be studied with simulations of split ring resonators[6] instead of a TE_{110} cavity.

Acknowledgments

This project is supported by the European Unions Horizon 2020 research and innovation program under the Marie Skodowska-Curie grant agreement No. 721559 and the Austrian Science Fund FWF, Doctoral Program No. W1252-N27. We also acknowledge Dr. Fritz Caspers for helping us with the design of microwave devices.

References

1. E. Widmann *et al.*, Hyperfine Interact. **240**, 5 (2019).
2. M.C. Simon, these proceedings.
3. N.F. Ramsey *et al.*, Rev. Mod. Phys. **62**, 541 (1990).
4. V.A. Kostelecký *et al.*, Phys. Rev. D **92**, 056002 (2015).
5. AC/DC Module User's Guide, Version 5.3, COMSOL, Inc., www.comsol.com.
6. W.N. Hardy *et al.*, Rev. Sci. Instrum. **52**, 213 (1981).

Prospects for Lorentz-Violation Searches
with Top Pair Production at the LHC and Future Colliders

A. Carle, N. Chanon, and S. Perriès

Université de Lyon, Université Claude Bernard Lyon 1,
CNRS-IN2P3, Institut de Physique Nucléaire de Lyon,
Villeurbanne 69622, France

This article presents prospects for Lorentz-violation searches with $t\bar{t}$ at the LHC and future colliders. After a short presentation of the Standard-Model Extension as a Lorentz-symmetry-breaking effective field theory, we will focus on $t\bar{t}$ production. We study the impact of Lorentz violation as a function of center-of-mass energy and evaluate the sensitivity of collider experiments to this signal.

1. Introduction

The top-quark sector of Standard-Model Extension (SME) is weakly constrained. Since the LHC is a top factory, it provides a unique opportunity to search for Lorentz violation (LV). The SME is an effective field theory including all LV operators. Here, we consider the LV CPT-even part of the lagrangian modifying the top-quark kinematics:[1]

$$\mathcal{L}^{\text{SME}} \supset \tfrac{i}{2}(c_L)_{\mu\nu}\bar{Q}_t\gamma^\mu \overset{\leftrightarrow}{D}{}^\nu Q_t + \tfrac{i}{2}(c_R)_{\mu\nu}\bar{U}_t\gamma^\mu \overset{\leftrightarrow}{D}{}^\nu U_t, \tag{1}$$

where Q_t and U_t denote the left- and right-handed top-quark spinors, respectively. The $c_{\mu\nu}$ coefficients are constant in an inertial frame, taken to be the Sun-centered frame. We aim at measuring the constant coefficients:[2]

$$c_{\mu\nu} = \tfrac{1}{2}\left[(c_L)_{\mu\nu} + (c_R)_{\mu\nu}\right], \qquad d_{\mu\nu} = \tfrac{1}{2}\left[(c_L)_{\mu\nu} - (c_R)_{\mu\nu}\right]. \tag{2}$$

Expressions for these coefficients in a laboratory frame on Earth will introduce a time dependence of the cross section for $t\bar{t}$ production owing to the Earth's rotation around its axis. This time dependence can be exploited to search for LV at hadron colliders.

To express the $c_{\mu\nu}$ coefficients in the reference frame of a hadron circular collider, we need:

• the latitude λ, i.e., the angle between the equator and the poles,

- the azimuth θ,[3] i.e., the angle between the Greenwich tangent vector and the clockwise ring collider tangent vector,
- the longitude impacts only the phase of the signal because of the Earth's rotation around its axis, and
- the Earth's angular velocity Ω.

2. Modulation of the $t\bar{t}$ cross section

The analysis aims at measuring the time dependence of the $t\bar{t}$ cross section

$$\sigma_{\text{SME}} = [1 + f(t)]\,\sigma_{\text{SM}}. \tag{3}$$

A first analysis of this kind was performed with the D0 detector at the Tevatron.[4] We use here the same benchmarks. We analyze Wilson's coefficients for a couple of non-null $c_{\mu\nu}$: $c_{XX} = c_{YY}$, $c_{XY} = -c_{YX}$, $c_{XZ} = c_{ZX}$ or $c_{YZ} = c_{ZY}$. Each of these scenarios generates an oscillating behavior of the amplitude. The latitude λ and the azimuth θ affect the amplitude while the Earth's angular velocity Ω affects the frequency. In the case of $c_{XX} = c_{YY}$ and $c_{XY} = -c_{YX}$, $f(t)$ has a period of one sidereal day. On the other hand, in the $c_{XZ} = c_{ZX}$ and $c_{YZ} = c_{ZY}$ case, the amplitude has a period of one half of a sideral day. More detailed expressions are given in Refs. 2,4.

3. Expected sensitivity

In this work, samples of $t\bar{t}$ with dilepton decay were generated with MadGraph-aMC@NLO 2.6. It was found that the amplitude of the LV $t\bar{t}$ signal is increasing with the center-of-mass energy. The signal amplitude as a function of the center-of-mass energy in p–p collisions (with CMS or ATLAS as the laboratory frame) increases from 0.001 at D0 (in the $c_{XY} = -c_{YX} = 0.01$ scenario) to 0.045 at the LHC Run II (13 TeV) and to 0.055 at the Future Circular Collider (FCC, 100 TeV).

We evaluate the expected sensitivity to the signal for each benchmark.[5] As a consequence of the increase in luminosity, the increase in cross section, and the increase in the amplitude of the LV signal, we find the following expected sensitivities to the SME coefficient $c_{\mu\nu}$ in the $c_{XX} = c_{YY}$ case:

- $\Delta c = 7 \times 10^{-1}$: D0 ($\sqrt{s} = 1.96$ TeV, $\mathcal{L} = 5.3$ fb^{-1}),
- $\Delta c = 1 \times 10^{-3}$: LHC Run II ($\sqrt{s} = 13$ TeV, $\mathcal{L} = 150$ fb^{-1}),
- $\Delta c = 2 \times 10^{-4}$: HL-LHC ($\sqrt{s} = 14$ TeV, $\mathcal{L} = 3000$ fb^{-1}),
- $\Delta c = 3 \times 10^{-5}$: LHC Run II ($\sqrt{s} = 27$ TeV, $\mathcal{L} = 15$ ab^{-1}),
- $\Delta c = 9 \times 10^{-6}$: LHC Run II ($\sqrt{s} = 100$ TeV, $\mathcal{L} = 15$ ab^{-1}).

206

4. Signal amplitude at hadron colliders

A noticeable fact is the dependence of the signal amplitude on the latitude and azimuth of the collider experiment on Earth. This dependence is presented in Fig. 1. We find that performing such an experiment at the

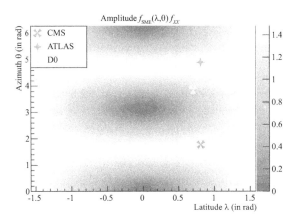

Fig. 1: Amplitude of $f(\lambda, \theta)$ as a function of latitude and azimuth for the XX, YY, and XY benchmarks.

LHC would increase the sensitivity to SME coefficients in the top sector by two orders of magnitude. Further improvements are expected at future colliders.

Acknowledgments

Thanks to Alan Kostelecký and Ralf Lehnert for giving me the opportunity to expose my work in front of a benevolent and hard-working community.

References

1. D. Colladay and V.A. Kostelecký, Phys. Rev. D **93**, 036005 (2016).
2. M.S. Berger, V.A. Kostelecký, and Z. Liu, Phys. Rev. D **93**, 036005 (2016).
3. M. Jones, Activity Report, EDMS, 322747 (2005).
4. D0 Collaboration, V.M. Abazov *et al.*, Phys. Rev. Lett. **108**, 261603 (2012).
5. A. Carle, N. Chanon, and S. Perriès, to appear.

Experimental Apparatus and Design for Parity-Odd Asymmetry Measurements in Compound Nuclei

C.J. Auton,* W.M. Snow,* J. Curole,* H. Lu,* G. Visser,* J. Doskow,*
J. Vanderwerp,* M. Gabel,* C. Crawford,† D. Schaper,† B. Plaster,† D. Olivera,†
D. Hajer,† L. Cole,† G. Forbes,† S.I. Penttilä,‡ T. Yamamoto,§ H.M. Shimizu,§
P. King,¶ L. Barrón-Palos,‖ and A. Martin‖

*Department of Physics, Indiana University, Bloomington, IN 47405, USA

†Department of Physics and Astronomy, University of Kentucky,
Lexington, KY 40506, USA

‡Oak Ridge National Laboratory, Oak Ridge, TN 37831, USA

§Department of Physics, Nagoya University,
Furo-cho, Chikusa-ku, Nagoya, 464-8601, Japan

¶Department of Physics and Astronomy, Ohio University, Athens, OH 45701, USA

‖Instituto de Física, Universidad Nacional Autónoma de México,
Apartado Postal 20-364, 01000, Mexico

The Neutron OPtics Time Reversal Experiment, or NOPTREX, collaboration plans to conduct a sensitive search for T violation in polarized neutron transmission through polarized compound nuclei by taking advantage of the very large amplification of symmetry-violating effects in p-wave resonances. As a step toward this experiment, we are remeasuring P violation in certain nuclei on resonance to greater precision. The first of these nuclei is ^{139}La with a large 10% P-odd asymmetry on the 0.734 eV p-wave resonance. This measurement is unique in that ^{139}La is used as both the spin filter and target. We aim for 1% accuracy on the 10% P-odd asymmetry at the resonance peak.

1. Introduction

Although originally thought to be fundamental symmetries, C, P, and CP symmetries are violated in various physical systems. The CPT theorem states that any Lorentz-invariant local quantum field theory with a hermitian hamiltonian must have CPT symmetry.[1,2] This implies that CP violation is equivalent to T violation. Observed CP violation can be explained in the Standard Model through two sources: the complex phase of the CKM matrix in the weak interaction, and the θ term in Quantum Chromodynamics. While the Standard Model can fully explain the CP violation seen in experiments, it underpredicts the observed size of the matter–antimatter asymmetry of the universe by several orders of magnitude if one believes the Big Bang theory arguments of Sakharov.[3] This implies that there are larger sources of CP violation outside of the Standard Model waiting to be discovered.

Searches for new sources of T violation have become increasingly important in nuclear, particle, and astrophysics. Theoretical ideas point to several possible types of new sources for T violation meaning that different searches for T violation cannot be equally sensitive to all mechanisms.[4–7] Thus, searches for T violation in different systems could lead to new discoveries. Since T violation has not been discovered in the nucleon sector, it is of fundamental importance to perform such experiments in any nuclear system. The required sensitivity for such T violation could be provided by p-wave neutron resonances in compound nuclei. The NOPTREX collaboration plans to conduct a sensitive search for T violation through an asymmetry in polarized-neutron transmission on epithermal p-wave resonances of polarized nuclei. Theory has suggested that T violation in compound nuclei has the potential to be two orders of magnitude more sensitive than the current limit from nEDM measurements given present spallation-neutron-source brightness.[8–12] The basic idea behind performing such a measurement is to look for a term in the neutron forward scattering amplitude of the form $\boldsymbol{\sigma}_n \cdot (\boldsymbol{k_n} \times \boldsymbol{I})$, where $\boldsymbol{\sigma}_n$ and $\boldsymbol{k_n}$ are the spin and momentum of the neutron, and $\boldsymbol{I}$ is the spin of the target nucleus. This observable is not only P-odd but also T-odd. Measuring an asymmetry in the forward scattering amplitude serves as a null test for time-reversal invariance and would in principle be insensitive to final-state effects.[8,9,13]

2. Experimental method and setup

The basic method for measuring the P-odd longitudinal asymmetry is to measure transmission through a target with two different neutron helicity states. This is done by passing unpolarized neutrons through a polarizer to give a nonzero longitudinal polarization to the beam. The neutrons then enter an adiabatic spin-flipper that can either flip the spin or leave it unchanged. This produces two helicity states, which strike the target nuclei. The transmission asymmetry is then measured using a neutron-sensitive scintillator viewed by a photomultiplier tube, which converts the instantaneous neutron current into an electric current and then a voltage. Lanthanum-139 is a good candidate target nucleus for searches of T violation because of the existing large 10% P violation on the 0.743 eV p-wave resonance. This large P violation actually weakly polarizes the neutron beam on resonance. In this case, ^{139}La can be used as both the polarizer and target.

Although this "Double Lanthanum" method has been done before, it was performed with the ^{139}La targets at room temperature.[14,15] The NOP-TREX Double Lanthanum experiment aims to increase the precision on the P-odd asymmetry by using cryogenic targets cooled to about 10 K. Doing this reduces the Doppler broadening on the resonance of interest.

Acknowledgments

This work is supported in part by the US National Science Foundation under grant No. PHY-1614545, the US Department of Education GAANN fellowship P200A150204, and the US Department of Energy. Additional funding from the Established Program to Stimulate Competitive Research as well as support from the APS Division of Nuclear Physics and the NSF Nuclear Physics Division under grant No. PHY-1812115 are gratefully acknowledged.

References

1. J. Schwinger, Phys. Rev. **82**, 914 (1951).
2. J. Schwinger, Phys. Rev. **91**, 713 (1953).
3. A. Sakharov, Pisma Zh. Eksp. Teor. Fifz. **5**, 32 (1967).
4. M. Pospelov, Phys. Lett. B **530**, 123 (2002).
5. M. Pospelov and A. Ritz, Ann. Phys. **318**, 119 (2005).
6. Y.-H. Song, R. Lazauskas, and V.P. Gudkov, Phys. Rev. C **84**, 025501 (2011).
7. Y.-H. Song, R. Lazauskas, and V.P. Gudkov, Phys. Rev. C **87**, 015501 (2013).
8. V.P. Gudkov and Y.-H. Song, Hyperfine Interact. **214**, 105 (2013).
9. J.D. Bowman and V.P. Gudkov, Phys. Rev. C **90**, 065503 (2014).
10. V.E. Bunakov and V.P. Gudkov, Nucl. Phys. A **401**, 93 (1983).
11. V.P. Gudkov, Phys. Rept. **212**, 77 (1992).
12. A.G. Beda and V.R. Skoy, Phys. Part. Nucl. **38**, 775 (2007).
13. V.P. Gudkov, Phys. Lett. B **243**, 319 (1990).
14. V. Yuan *et al.*, Phys. Rev. C **44**, 2187 (1991).
15. C.D. Bowman, J.D. Bowman, and V.W. Yuan, Phys. Rev. C **39**, 1721 (1989).

The MICROSCOPE Space Mission and Lorentz Violation

Geoffrey Mo,[*] Hélène Pihan-Le Bars,[†] Quentin G. Bailey,[‡] Christine Guerlin,[†,§]
Jay D. Tasson,[*] and Peter Wolf[†]

[*]*Department of Physics and Astronomy, Carleton College,*
Northfield, MN 55057, USA

[†]*SYRTE, Observatoire de Paris, Université PSL, CNRS,*
Sorbonne Université, LNE, 75014 Paris, France

[‡]*Department of Physics and Astronomy, Embry-Riddle Aeronautical University,*
Prescott, AZ 86301, USA

[§]*Laboratoire Kastler Brossel, ENS-Université PSL, CNRS, Sorbonne Université,*
Collège de France, 24 rue Lhomond, 75005 Paris, France

In this contribution to the CPT'19 proceedings, we summarize efforts that use
data from the MICROSCOPE space mission to search for Lorentz violation in
the Standard-Model Extension.

1. The SME and the Weak Equivalence Principle

Data from the MICROSCOPE space mission can be used to search for
Lorentz violation within the field-theoretic framework of the Standard-
Model Extension (SME).[1] The SME can roughly be thought of as a series expansion about known physics at the level of the action that forms
a broad and general framework for tests of Lorentz symmetry.[2,3] Terms in
the SME action are constructed from Lorentz-violating operators along with
coefficients for Lorentz violation that characterize the amount of Lorentz
violation in the theory. In general, the coefficients for Lorentz violation
are particle-species dependent such that in the fermion sector couplings to
Lorentz violation may differ for protons, neutrons, and electrons. When
couplings to gravity are considered in the fermion sector, this species dependence leads to effective Weak Equivalence Principle (WEP) violations.[4]

In the present context, we consider a single coefficient field $(a_{\text{eff}})_\mu$ in
the classical point-particle limit, where the matter-sector action is[4]

$$S_U^{\text{B}} \approx \int d\lambda \big(-m^{\text{B}} \sqrt{-g_{\mu\nu} u^\mu u^\nu} - (a_{\text{eff}}^{\text{B}})_\mu u^\mu \big), \tag{1}$$

for a body B of mass m^{B} with the post-Newtonian metric $g_{\mu\nu}$ and four-velocity u^μ. In spontaneous Lorentz-violation models, $(a_{\mathrm{eff}})_\mu$ develops a vacuum expectation value $(\bar{a}_{\mathrm{eff}})_\mu$. An additional freedom associated with a coupling constant in models of spontaneous breaking is characterized by α.[4]

When these coefficients are taken into account in a Newtonian example via the action (1), the equations of motion for bodies become

$$m^{\mathrm{B}}\vec{a} = (m^{\mathrm{B}} + 2\alpha(\bar{a}_{\mathrm{eff}}^{\mathrm{B}})_t)\vec{g}, \tag{2}$$

where $\vec{a}$ is the acceleration, $\vec{g}$ is the Newtonian gravitational field, and $(\bar{a}_{\mathrm{eff}}^{\mathrm{B}})_\mu = \sum_w N_{\mathrm{B}}^w (\bar{a}_{\mathrm{eff}}^w)_\mu$.[4] Here, the sum is over particle species $w = $ proton, neutron, electron, and N_{B}^w is the number of particles of type w in the body.

The WEP states that the gravitational mass m_{grav} is equal to the inertial mass m_{inert}.[5] In other words,

$$m_{\mathrm{inert}}\vec{a} = m_{\mathrm{grav}}\vec{g} \tag{3}$$

can be rewritten as $\vec{a} = \vec{g}$. Hence, the signal in a WEP experiment is a relative acceleration of a pair of co-located bodies 1 and 2 in free fall in a gravitational field. Traditional constraints on Lorentz-invariant WEP violations have been quantified by the Eötvös parameter $\delta(A, B) = 2(\mathrm{a}_1 - \mathrm{a}_2)/(\mathrm{a}_1 + \mathrm{a}_2)$, where a_B is the free-fall acceleration of the body.

Comparison of Eqs. (2) and (3) reveals the effective WEP violation induced by $(\bar{a}_{\mathrm{eff}})_\mu$. However, the Lorentz-violation signal is more complicated and cannot be characterized by a single parameter such as δ. This can be seen already from the Newtonian result, where the time component of the coefficient for Lorentz violation will mix with the spatial components under a time-dependent boost, such as the boost of the Earth around the Sun, yielding a time-dependent WEP signal involving multiple components of $(\bar{a}_{\mathrm{eff}})_\mu$. When the relative free-fall rate for a pair of electrically neutral bodies 1 and 2 is considered, the signals are proportional to[4]

$$\sum_w \left(\frac{N_1^w}{m^1} - \frac{N_2^w}{m^2} \right) (\bar{a}_{\mathrm{eff}}^w)_\mu \approx \frac{N_1^n N_2^p - N_1^p N_2^n}{m^1 m^2} m^n (\bar{a}_{\mathrm{eff}}^{n-e-p})_\mu \equiv A(\bar{a}_{\mathrm{eff}}^{n-e-p})_\mu, \tag{4}$$

where $(\bar{a}_{\mathrm{eff}}^{n-e-p})_\mu \equiv (\bar{a}_{\mathrm{eff}}^n)_\mu - (\bar{a}_{\mathrm{eff}}^e)_\mu - (\bar{a}_{\mathrm{eff}}^p)_\mu$.

2. MICROSCOPE space mission

The MICROSCOPE space mission is a French CNES microsatellite designed to test the WEP with the best-ever precision of one part in 10^{15}.[6] It was launched in 2016 and completed its data-taking in 2018. It contains an instrument composed of two concentric cylindrical test masses made of the platinum alloy Pt:Rh (90:10) and the titanium alloy Ti:Al:V (90:6:4) for the inner and outer mass, respectively. With these measurements, the MICROSCOPE team found a constraint on the Eötvös parameter of $\delta(\text{Ti, Pt}) = [1 \pm 9 \text{ (stat)} \pm 9 \text{ (syst)}] \times 10^{-15}$, representing an improvement of over an order of magnitude over the previous best limits.[7]

A subset of the Lorentz-violation signals arise at frequencies different from the WEP signal and may remain hidden in the analysis aimed at the traditional WEP. Hence, a separate analysis has been performed to search for signals at the additional frequencies associated with $(\bar{a}_{\text{eff}}^{n-e-p})_\mu$.[1] Constraints on coefficients were analyzed globally treating all four components together as well as for "maximal reach,"[9] where only one coefficient was assumed to be nonzero at a time. With the alloys used for the test bodies, the parameter A in Eq. (4) is about $0.06\,\text{GeV}^{-1}$. With this, improvements on prior sensitivities[8] of about one to two orders of magnitude were achieved.[1]

Acknowledgments

GM is grateful for support from the Carleton College Towsley fund, and QGB was supported by US National Science Foundation grant 1806871.

References

1. Q.G. Bailey, C. Guerlin, G. Mo, H. Pihan-Le Bars, J.D. Tasson, and P. Wolf, in preparation.
2. D. Colladay and V.A. Kostelecký, Phys. Rev. D **58**, 116002 (1998); V.A. Kostelecký, Phys. Rev. D **69**, 105009 (2004).
3. For a review, see J.D. Tasson, Rep. Prog. Phys. **77**, 062901 (2014).
4. V.A. Kostelecký and J.D. Tasson, Phys. Rev. D **83**, 016013 (2011).
5. C.M. Will, *Theory and Experiment in Gravitational Physics*, Cambridge University Press, Cambridge, 1993.
6. P. Touboul *et al.*, Comptes Rendus Acad. Sci. Série IV **2**, 9 (2001).
7. P. Touboul *et al.*, Phys. Rev. Lett. **119**, 231101 (2017).
8. *Data Tables for Lorentz and CPT Violation*, V.A. Kostelecký and N. Russell, 2019 edition, arXiv:0801.0287v12; A. Bourgoin *et al.*, Phys. Rev. Lett. **119**, 201102 (2017).
9. N.A. Flowers, C. Goodge, and J.D. Tasson, Phys. Rev. Lett. **119**, 201101 (2017).

Exotic Spin-Dependent Interaction Searches at Indiana University

I. Lee, J. Shortino, J. Biermen, A. Din, A. Grossman, M. Gabel, E. Guess, C.-Y. Liu,
J.C. Long, S. Reger, A. Reid, M. Severinov, B. Short, W.M. Snow, E. Smith,
and M. Zhang

Department of Physics, Indiana University, Bloomington, IN 47405, USA

On behalf of the ARIADNE Collaboration

The axion is a hypothesized particle appearing in various theories beyond the Standard Model. It is a light spin-0 boson initially postulated to solve the strong CP problem and is also a strong candidate for dark matter. If the axion or an axion-like particle exists, it would mediate a P-odd and T-odd spin-dependent interaction. We describe two experiments under development at Indiana University Bloomington to search for such an interaction.

1. Introduction

The Standard Model possesses many unexplained features. Why QCD does not violate CP is one of them. Peccei and Quinn in 1977 proposed that CP conservation in the strong interactions can be explained by axions.[1] Axions appear in many other beyond-the-Standard-Model theories, including string theory,[2] and it is also considered as a strong candidate for dark matter.[3]

The axion leads to a P- and T-odd spin-dependent potential accessible in laboratory tests.[4,5] Two new experiments with improved sensitivities are being developed. The PTB experiment exploits the world-renowned magnetically shielded room at Physikalisch-Technische Bundesanstalt (PTB) in Germany along with their highly developed SQUID-magnetometry technology. The Axion Resonant InterAction DetectioN Experiment (ARIADNE) is a collaboration among institutions in Korea, Canada, and the US.

2. Theory

The axion would mediate a short-range monopole-dipole interaction with a potential of the form

$$U(r) = \frac{\hbar^2 g_s g_p}{8\pi m_f} \left(\frac{1}{r\lambda_a} + \frac{1}{r^2} \right) e^{-(r/\lambda_a)} (\hat{\sigma} \cdot \hat{r}), \tag{1}$$

where g_s and g_p are coupling constants, m_f is the fermion mass, $\hat{\sigma}$ is the Pauli spin matrix, r is the distance between fermions, and $\lambda_a = h/m_a c$ is the axion Compton wavelength.[6] The interaction has a Yukawa-like potential, so its strength drops quickly beyond the axion Compton wavelength. It affects the spin of dipole particles as the magnetic field does, but since it is mediated by the axion, magnetic shielding has no effect on it.

3. PTB experiment

The PTB experiment features a rotating disk with segments of alternating materials having similar magnetic properties but different nucleon densities, hence providing a time-varying axion field. A cell with hyperpolarized ^{3}He and ^{129}Xe will be placed near the mass, positioned to experience the axion field from only one material at a time. The axion field perpendicular to the longitudinal polarization axis of the samples causes the precession.

The resonant amplification of the signal gained by matching the frequency of the time-varying axion potential to the precession frequency of the hyperpolarized nuclei can greatly improve the sensitivity compared to previous experiments. The 8-layered magnetically shielded room at PTB[7] allows a long spin relaxation time of the polarized gas samples on the order of 10^4 s to accumulate the resonant amplification effect. The transverse magnetization of precessing samples will be measured by PTB's highly sensitive SQUID. The mixture of ^{3}He and ^{129}Xe can distinguish axion-mediated interaction from magnetic effects by treating one of the species as a comagnetometer. Nevertheless, we suspect that magnetic impurities in the test masses will eventually pose a fundamental limitation on the sensitivity of this approach.

4. ARIADNE basics

The experimental principles used in the PTB experiment are also applied to the ARIADNE experiment. However, the ARIADNE experiment will be conducted at $4\,$K in a liquid-helium cryostat, so that superconducting magnetic shielding can be employed to eliminate any test-mass magnetic-impurity systematics. The source mass is a rotating tungsten sprocket with 22 alternating segments. Three quartz blocks, each having a sample cell, bias coils, a SQUID loop, and niobium coating, will be placed next to the tungsten source mass. The blocks are thermally anchored to a $4\,$K copper plate turning its niobium coating into a superconducting shield against external magnetic backgrounds. The hyperpolarized ^{3}He samples get cooled to 4 K increasing the density to improve the signal level. Such advantages of conducting the experiment at 4 K add up to several orders of magnitude improvement in sensitivity.

5. ARIADNE at Indiana University

The ^{3}He gas for ARIADNE is hyperpolarized by metastability exchange optical pumping. Indiana University will provide the hyperpolarized ^{3}He at 4 K. A recycled liquid-helium cryostat has been modified to house a Pyrex glass cell in which the entire process of polarizing and cooling ^{3}He gas occurs. A complete metastability-exchange-optical-pumping system will be installed above the cryostat, and the polarized gas will diffuse down to the test cell into the 4 K region. An NMR coil surrounding the test cell will be used to study the behavior of the polarized ^{3}He samples at 4 K.

An RF-shielded room is under construction at Indiana University on a vibrationally isolated floor in the subbasement of the physics building. The floor is covered by RF shielding ferrite tiles and fine copper mesh. A Faraday cage with copper mesh will surround the experimental apparatus. This could be a possible experimental site for the ARIADNE project.

Acknowledgments

We acknowledge support from the US National Science Foundation under award Nos. NSF PHY-1509176, NSF PHY-1509805, NSF PHY-1806395, NSF PHY-1806671, and NSF PHY-1806757.

References

1. R.D. Peccei and H.R. Quinn, Phys. Rev. Lett. **38**, 1440 (1977).
2. A. Arvanitaki *et al.*, Phys. Rev. D **81**, 123530 (2010).
3. G.R. Blumenthal *et al.*, Nature **311**, 517 (1984).
4. M. Bulatowicz *et al.*, Phys. Rev. Lett. **111**, 102001 (2013).
5. P.H. Chu *et al.*, Phys. Rev. D **87**, 011105 (2013).
6. J.E. Moody and F. Wilczek, Phys. Rev. D **30**, 130 (1984).
7. J. Bork, H.D. Hahlbohm, R. Klein, and A. Schnabel, in J. Nenonen, R.J. Ilmoniemi, and T. Katila, eds., *Biomag2000, Proc. 12th Int. Conf. on Biomagnetism*, Helsinki University of Technology, Espoo, Finland, 2001.

Highlights from the Seven-Year
High Energy Starting Events Sample in IceCube

Kareem Farrag[*,†]

*Queen Mary University of London, E1 4NS, UK

†University of Southampton, Southampton, SO17 1BJ, UK

On behalf of the IceCube Collaboration

Here, we outline the main highlights from the seven-year High Energy Starting Events event sample. The next new-physics search using astrophysical neutrino flavor data is described, where we reach the Planck scale for the first time.

The High Energy Starting Events (HESE) data sample in IceCube[1] is selected to extract a sample of astrophysical neutrinos with high purity.[2,3] Events are accepted into the sample if the interaction vertex is contained in a subvolume of IceCube defined by the inner part of the detector as the outermost layers are used as a veto region to reject atmospheric backgrounds, as shown in Fig. 1.[4] We find 102 events observed over 2635 days with 60 events above 60 TeV in deposited energy. HESE is a low-atmospheric-background event selection used in IceCube to study astrophysical neutrinos including dark-matter searches and anomalous spacetime effects through the astrophysical neutrino flavour composition.

Key changes in the event reconstruction include global changes to the ice model. This includes both ice anisotropy and tilt effects. HESE 7 is consistent with a single power-law fit with spectral index $\gamma \sim 2.9$. Systematic uncertainties include contributions from atmospheric neutrino fluxes as well as detector systematics. We show the energy and angular distributions in Fig. 2. The preliminary best fit for the flavor composition of diffuse neutrinos is $(0.29 : 0.50 : 0.21)$, where the current flavor contour is consistent with $(1 : 1 : 1)$, see Fig. 3. Note that zero tau events cannot be ruled out with the best-fit $E^{-2.9}$ spectrum. Contours are computed with Wilk's theorem, and work along this avenue is ongoing. A Beyond-the-Standard-Model search using the Standard-Model Extension[5] was performed attempting to detect anomalous flavor ratios. Such ratios could arise due to the presence of effective operators at high energy scales. Three source flavor compositions of the

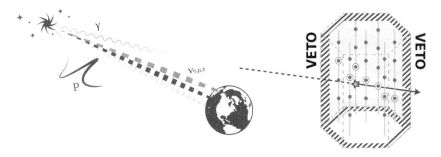

Fig. 1: Left: Illustration of astrophysical neutrino interferometry. Neutrinos at production and detection exist in flavor eigenstates. However, they travel in the so-called propagation basis. This basis evolves with time according to their effective hamiltonian. Right: Illustration of the HESE selection. The trigger demands that the interaction vertex be fully contained (bounded by the dashed region). Events require more than 6000 photoelectrons to ensure (to 99.999%) that cosmic-ray muons would produce enough light in the veto region to be excluded.[2]

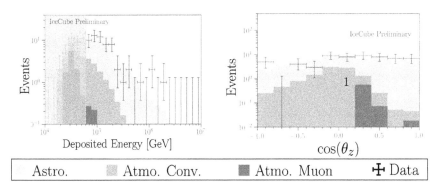

Fig. 2: Left: HESE 7 energy distribution. The x axis shows the deposited-energy estimate by a neutrino-event interaction. Atmospheric conventional and muon estimates in the sample are the mid- and dark-gray regions, respectively. The astrophysical component in light gray has a harder energy spectrum with spectral index $\gamma \sim 2.9$. The number of events in each bin are marked by a cross. Right: Plot of the zenith distribution for the HESE sample.

form $(f_e : f_\mu : f_\tau)$ are believed to dominate, namely $(1 : 2 : 0)$, $(1 : 0 : 0)$, and $(0 : 1 : 0)$. We expect to place the most stringent limits on higher-order new-physics operators. Furthermore, two tau candidate neutrinos have been observed in HESE, with double cascade energies $E \sim 100\,\mathrm{TeV}$ and $1.8\,\mathrm{PeV}$. This corresponds to the first astrophysical tau-neutrino candidates in IceCube. Current work is ongoing to quantify their significance.

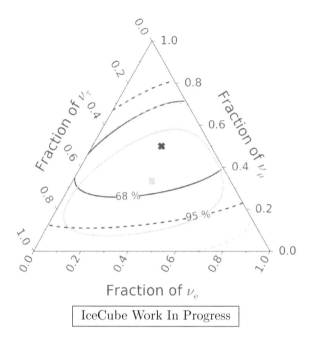

Fraction of ν_e

IceCube Work In Progress

Fig. 3: Ternary plot of current sensitivity and data contour. The data are consistent with a (1 : 1 : 1) flavor ratio on Earth. The gray contour shows sensitivity with best fit crossed at (1 : 1 : 1). Black contour shows data contour for HESE with ternary ID topology with cross at best fit (0.29 : 0.50 : 0.21). Solid and dashed lines show 68% and 95% credibility regions, respectively.

In conclusion, our first-ever search for Lorentz violation was done using astrophysical neutrinos. Most notably, our limit resides several orders of magnitude below the Planck scale. More work is needed to constrain all source flavor paradigms. Our technique reaches the quantum-gravity regime for the first time. Future searches will be able to probe the region of flavor space currently inaccessible with seven years of HESE data.

References

1. IceCube Collaboration, M.G. Aartsen *et al.*, J. Instrum. **12**, P03012 (2017).
2. IceCube Collaboration, M.G. Aartsen *et al.*, Science **342**, 1242856 (2013).
3. IceCube Collaboration, J. Stachurska Eur. Phys. J. Web Conf. **207**, 02005 (2019).
4. C.A. Argüelles *et al.*, J. Cosmol. Astropart. Phys. **1807**, 047 (2018).
5. V.A. Kostelecký, Phys. Rev. D **69**, 105009 (2004).

Testing for Lorentz-Invariance Violations Through Birefringence Effects on Gravitational Waves

K. O'Neal-Ault, Quentin G. Bailey, and M. Zanolin

Department of Physics and Astronomy, Embry-Riddle Aeronautical University,
Prescott, AZ 86301, USA

LIGO-VIRGO gravitational-wave data provide a new means of testing for
Lorentz violation; a methodology is developed to test for Lorentz violation
using gravitational wave-data via the PIRO Simulation. We focus on the gravity sector of the Standard-Model Extension, where a possible observable effect
is birefringence: we would observe a small shift in the arrival times between
the two gravitational-wave polarizations from a distant source at the detectors.

1. Background

We focus on gravitational-wave (GW) data from the LIGO interferometers
that will be used to constrain coefficients from the SME. The weak-field
limit is implemented, where $g_{\mu\nu} = \eta_{\mu\nu} + h_{\mu\nu}$, and $h_{\mu\nu}$ is considered as a
small perturbation around the flat Minkowski spacetime metric $\eta_{\mu\nu}$.

If there is a background field that couples to gravity, then Lorentz symmetry can be broken.[1] GWs will couple differently depending upon their
orientation with respect to these fields producing different dynamical effects. They have two independent polarizations in the TT-gauge, h_+ and
$h_\times$ (this remains true to leading order in LV). There are limits that already
exist for some coefficients, which are listed in a reference of data tables.[2,3]

A possible effect is birefringence leading a delay in arrival times at the
LIGO detectors between the two polarizations. This delay is given by[4]

$$\Delta t \approx \sum_{d=5} 2\omega^{d-4} \int_0^z \frac{(1+z)^{d-4}}{H_z} dz \sum_{jm} Y_{jm}(\hat{n}) \mathcal{K}^{(d)}_{(v)jm}. \tag{1}$$

The time delay is written as an expansion of spherical harmonics, Y_{jm},
where $\hat{n}$ is in the direction of propagation of the GW. The $\mathcal{K}^{(d)}_{(v)jm}$ are the
background fields or coefficients assumed small and constant. The distance
is calculated in terms of the redshift and the cosmological luminosity distance, and ω is the frequency.

2. Initial setup and code development

A χ^2 test is used to test for birefringence. It indicates which time shift in the model best matches that in the strain data. An extreme emission model, PIRO, is used to develop the methodology for the analysis.[5] The model is of a failed supernova with a black hole in the center of a disk of unstable, inspiraling matter.

A model strain and a pseudo data strain were generated,

$$S^A_{\mathrm{model}}(t) = F^A_+(\theta, \phi, \psi) \, h_+(\theta, \phi, \psi, t, \Delta t) + F^A_\times(\theta, \phi, \psi) \, h_\times(\theta, \phi, \psi, t), \quad (2)$$

where $F^A_+(\theta, \phi, \psi)$ and $F^A_\times(\theta, \phi, \psi)$ are the plus and cross antenna response functions for either Livingston or Hanford. The Δt represents the time shift in the plus polarization due to the LV background field. The pseudo strain data was formed by adding LIGO noise to the PIRO strain.

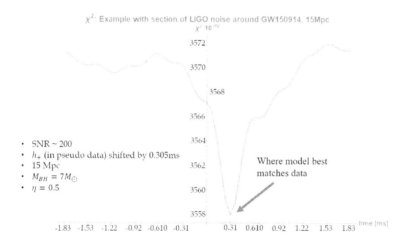

Fig. 1: χ^2 test for PIRO WF at 10 kpc with generated, scaled noise of an SNR of 200. Initially, a PIRO waveform of a seven-solar-mass black hole with the infalling matter mass at a fraction of this, namely $\eta = 0.5$, and a face-on orientation with respect to the LIGO detectors at a distance of 10 kpc was used. Also considered were waveforms at 1 Mpc and 1 Gpc distances with all other parameters the same.

The minimum value of the χ^2 test (Fig. 1) shows where, if any, the time shift between polarizations occurs in the pseudo data strain. The resolvability of the time shift will depend on the SNR and the resolution of the other source parameters.

To build confidence in the result, a histogram is used. Each run with a different implemented noise calculates the χ^2 minimum. The spread in distribution is related to the SNR, where a larger spread correlates to a lower SNR. The histogram peak would indicate a Δt between polarizations.

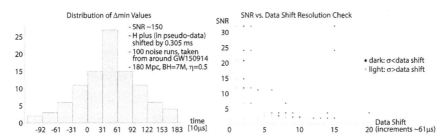

Fig. 2: Left: Histogram distribution of χ^2 mininum values. They can help provide limits on SME coefficients on the left. This does not include frequency-dependent effects. Right: Confidence level and ability to resolve LV effects across different SNR values.

To better understand the relationship between different SNR values and the ability to resolve the LV time shift between polarizations confidently, a plot is generated of SNR versus Δt (Fig. 2). For a given SNR, we can compare one standard deviation (1σ) to the mean LV shift of the histogram. We would expect it easier to resolve the Δt if it is larger than 1σ. If $\Delta t < 1\sigma$, a constraint can still be placed on the SME coefficients. Also noting that the LV coefficients depend on the source's location on the sky, and given many events, one can potentially map out the background fields across the sky. One can also include the effects of other parameters using a fitting-factor technique.[6]

References

1. Q.G. Bailey and V.A. Kostelecký, Phys. Rev. D **74**, 045001 (2006).
2. *Data Tables for Lorentz and CPT Violation,* V.A. Kostelecký and N. Russell, 2019 edition, arXiv:0801.0287v12.
3. V.A. Kostelecký and M. Mewes, Phys. Lett. B **779**, 136 (2018).
4. V.A. Kostelecký and M. Mewes, Phys. Lett. B **757**, 510 (2016).
5. A.L. Piro and E. Thrane, Astrophys. J. **761**, 63 (2012).
6. R. Tso and M. Zanolin, Phys. Rev. D **93**, 124033 (2016).

Neutron Spin Rotation: Exotic Spin-Dependent Interactions

Kyle Steffen

Department of Physics, Indiana University,
Bloomington, IN 47405, USA

On behalf of the NSR Collaboration

Exotic interactions between fermions are predicted by many beyond-the-Standard-Model theories, several of which can be viewed through exotic spin-1 boson exchange. An experiment in search of axial-vector couplings to this field was performed using a slow-neutron polarimeter containing a slotted rotatable target system designed with atom-density gradients that would tip the plane of polarization into the neutron momentum. For the V_5 potential, the upper bound on an exotic axial neutron–matter coupling constant g_A^2 was improved by about three orders of magnitude for mm–μm range interactions, and a new target design gives a projected two orders of magnitude further improvement. A related spin-echo measurement technique allows for measurement of $g_V g_A$ interactions and has access to spin–gravity couplings in the matter sector of the SME.

1. Motivation

Parallel to the context of the SME, many experiments that search for new physics can be characterized in the formalism of long-range interactions between fermions that are mediated by light exotic exchange bosons.[1] A general parameterization of the Lagrange density for spin-1 boson X^μ,

$$\mathcal{L} = \bar{\psi}(g_V \gamma^\mu + g_A \gamma^\mu \gamma^5)\psi X_\mu, \tag{1}$$

results in an axial-vector non-relativistic interaction potential, which was probed by Piegesa and Pignol using Ramsey's technique of oscillatory fields to search for pseudomagnetic precession caused by a sample plate parallel to a neutron beam.[2] The potential is of the form

$$V_{\text{Axial}}(\vec{r}, \vec{v}) \equiv V_5 = \frac{g_A^2}{4\pi m_n} \frac{e^{-m_0 r}}{r} \left(\frac{1}{r} + \frac{1}{\lambda_c} \right) \vec{\sigma} \cdot (\vec{v} \times \hat{r}), \tag{2}$$

where m_n is the mass of the neutron, and $m_0 = \frac{1}{\lambda_c}$ is the mass of the exchange boson.

2. Measurement method and apparatus

By using a slow-neutron polarimeter, the pseudomagnetic precession of a polarized neutron's magnetic moment induced via V_5 by a nearby fermion density gradient can be measured as a change in transmitted intensity through an analyzer.[3] The NSR polarimeter, constructed to measure parity violation in n–^{4}He,[4,5] contains a segmented neutron guide system allowing for two parallel measurements to occur simultaneously. To further suppress systematics, each parallel sub-beam may be further split by the target into a sample beam and a reference beam, which may be switched by discrete rotations of the target. The ^{3}He ion detector used is segmented into transverse quadrants for individual measurements of each sub-beam and longitudinally segmented to discriminate wavelength-dependent systematics.

3. Results

The NSR apparatus was installed on flight path 12 at the Los Alamos Neutron Science Center in 2016, and the experiment was performed with one week of beam time. The measured rotation of $\varphi' = [2.8 \pm 4.6 \text{(stat.)} \pm 4.0 \text{(sys.)}] \times 10^{-5} \, \text{rad/m}$ was consistent with zero, but yielded limits on g_A^2 in the mm–μm range,[6] as seen in Fig. 1.

4. Upcoming measurements

To further constrain g_A^2 with this apparatus, the target should be modified to increase the expected V_5 signal, while further suppressing systematics to warrant operation on a continuous high-intensity neutron beamline, such as NG-C at NIST. To this end, we are creating a geometrically identical version of the target with the copper source plates replaced with electropolished tungsten plates for a larger density gradient resulting in an increased V_5 signal. Also, a target without density gradient should act as a null test of the apparatus. All target plates will undergo precise magnetic mapping with a commercial rubidium zero-field magnetometer. The projected sensitivity of this measurement is also shown in Fig. 1. In a related experiment, the OffSpec neutron spin-echo instrument at ISIS may be used to search for both g_A^2 and $g_V g_A$ couplings simultaneously by using the Earth as a source mass. This instrument has been used previously to measure gravitationally induced phase shifts by operating as a neutron gravitational interferometer.[7] The spin-echo technique splits neutron spin eigenstates in space allowing them to independently accumulate phase before being recombined before an analyzer. The ability of OffSpec to rotate the spin-echo splitting plane from horizontal to vertical and tilt to multiple inclination angles also allows the search for spin–gravity couplings within the SME matter sector.[a]

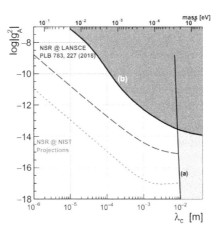

Fig. 1: Our experimental constraints and projections on g_A^2 as a function of λ_c compared with results from (a) K–^{3}He comagnetometry,[8] and from (b) Ramsey spectroscopy.[2]

Acknowledgments

This work is supported by the US National Science Foundation under grant No. PHY-1614545.

References

1. B. Dobrescu and I. Mocioiu, J. High Energy Phys. **0611**, 005 (2006).
2. F.M. Piegsa and G. Pignol, Phys. Rev. Lett. **108**, 181801 (2012).
3. C. Haddock *et al.*, Nucl. Instrum. Meth. Phys. Res. A **885**, 105 (2018).
4. W.M. Snow *et al.*, Rev. Sci. Instrum. **86**, 055101 (2015).
5. H.E. Swanson *et al.*, accepted for publication in Phys. Rev. C.
6. C. Haddock *et al.*, Phys. Lett. B **783**, 227 (2018).
7. V.-O. de Haan *et al.*, Phys. Rev. A **89**, 063611 (2014).
8. G. Vasilakis, J.M. Brown, T.W. Kornack, and M.V. Romalis, Phys. Rev. Lett. **103**, 261801 (2009).

---

[a]The OffSpec spin–gravity experiment is being performed by a yet unnamed collaboration comprised of Robert Dalgliesh, Niels Geerits, Steven Parnell, Jeroen Plomp, Roger Pynn, W. Michael Snow, Kyle Steffen, Nina-Juliane Steinke, Ad van Well, and Victor de Haan in July 2019.

Ring Laser Gyroscope Tests of Lorentz Symmetry

Max L. Trostel,* Serena Moseley,* Nicholas Scaramuzza,[†] and Jay D. Tasson*

*Physics and Astronomy Department, Carleton College, Northfield, MN 55057, USA

[†]Physics Department, St. Olaf College, Northfield, MN 55057, USA

Interferometric gyroscope systems are being developed with the goal of measuring general-relativistic effects including frame-dragging effects. Such devices
are also capable of performing searches for Lorentz violation. We summarize
efforts that relate gyroscope measurements to coefficients for Lorentz violation
in the gravity sector of the Standard-Model Extension.

1. Interferometric gyroscopes and Lorentz violation

Lorentz violation in the Standard-Model Extension (SME) can be sought[1]
using interferometric gyroscopes based on light[2] or matter[3] waves. Here,
we summarize some results from Ref. 1 with a focus on light-based systems,
which consist of beams traveling around a closed path in opposite directions.
Effects that break the symmetry of the counter-propagating beams are
encoded in their interference. The largest such effect routinely observed,
the Sagnac effect, is due to rotation, which generates a beat frequency
$\nu_s = \frac{4A\vec{\Omega}\cdot\hat{n}}{P\lambda}$, where A is the area enclosed, $\vec{\Omega}$ is the angular velocity, $\hat{n}$ is
the vector normal to the loop, P is its perimeter, and λ is the wavelength
of the light. A fixed system on Earth will experience several effects that
alter the beat frequency, including the Sagnac effect of Earth's rotation
at angular frequency ω and the general-relativistic frame-dragging effect.
Lorentz violation as described by the SME[4,5] can also break the symmetry.

The contributions to a post-Newtonian expansion of the metric $g_{\mu\nu}$ in
the SME[6] that are relevant for our analysis take the form

$$g_{0j} = -\overline{s}^{0j}U - \overline{s}^{0k}U^{jk} + \tfrac{1}{2}\hat{Q}^j\chi, \tag{1}$$

where U is the Newtonian potential, χ is the superpotential,[6] U^{jk} is an
additional post-Newtonian potential,[6] and $\overline{s}^{\mu\nu}$ is the $d = 4$ coefficient for
Lorentz violation that provides the relevant minimal effects. The $d = 5$
effects are contained in $\hat{Q}^j = [q^{(5)0jk0l0m} + q^{(5)n0knljm} + q^{(5)njknl0m}]\partial_k\partial_l\partial_m$
where $q^{(5)\mu\rho\alpha\nu\beta\sigma\gamma}$ is the coefficient for Lorentz violation.[7]

2. Measuring Lorentz violation

The beat frequency measured by a light-based gyroscope[8] is related to the proper time difference $\Delta\tau$ between the paths taken by the counter-propagating beams via $\nu = \frac{\Delta\tau}{\lambda P}$, and at leading order in Lorentz violation we find in our post-Newtonian expansion:

$$\Delta\tau \approx 2 \oint g_{0j}dx^j, \qquad (2)$$

where the integral is taken around the closed interferometer loop. By analogy with Ampère's law, we transform this line integral into an integration over area, which simplifies into a product. In order to derive a general result that is valid for any Earth-based laboratory and for any ring orientation within that laboratory, we use the angles θ, ϕ, α, and β shown in Fig. 1. Given $d = 4$ coefficients for Lorentz violation, we find a beat frequency[1]

$$\nu_{LV}^{(4)} = \frac{4AGM}{\lambda PR^2}\sin\alpha[\cos\beta(\overline{s}^{TX}\sin\phi - \overline{s}^{TY}\cos\phi)$$
$$+ \sin\beta(\cos\theta(\overline{s}^{TX}\cos\phi + \overline{s}^{TY}\sin\phi) - \overline{s}^{TZ}\sin\theta)], \qquad (3)$$

where M and R are Earth's mass and radius, respectively. The small terms suppressed by an Earth-revolution boost factor have been omitted. A sample special case of this result is found in Ref. 9. For $d = 5$ coefficients, we find a beat frequency with a similar form (omitted here for brevity), which can be written in terms of the 15 canonical coefficient combinations written as components of K_{JKLM}.[10] The angle ϕ varies at the sidereal frequency, which indicates that both dimension four and five coefficients for Lorentz violation will produce signals with this time dependence.

3. Experiments

A number of planned or ongoing experiments may be of interest in the context of this work. For example, the Gyroscopes IN GEneral Relativity (GINGER) experiment, which is designed to measure the de Sitter and Lense–Thirring effects of General Relativity,[8,11] expects to obtain sensitivities to the angular velocity of the Earth via the Sagnac effect beyond the part in 10^9 level reaching perhaps a few parts in 10^{12}.[11] It is possible to generate crude estimates of the sensitivity to Lorentz violation that might be expected using these goals and the results outlined in Sec. 2. We find $\overline{s}^{TJ} \approx \frac{\epsilon\omega cR^2}{GM} \approx 10^{-6}$ and $K_{JKLM} \approx \frac{\epsilon\omega cR^3}{GM} \approx 10$ m in SI units with $\epsilon = 10^{-9}$ as the fractional sensitivity to ω and c as the speed of light. For $\overline{s}^{TJ}$, these sensitivities are competitive with other laboratory experiments,[12] while for

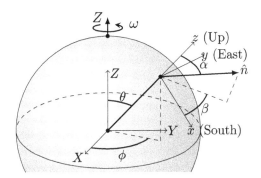

Fig. 1: Location of the laboratory in shifted Sun-centered frame[12] axes (X, Y, Z) and orientation of the normal vector of the gyroscope $\hat{n}$ relative to laboratory coordinates (x, y, z).

dimension five coefficients, these sensitivities are competitive with the best existing measurements from binary pulsars.[10]

Acknowledgments

The authors gratefully acknowledge financial support as follows: SM from Carleton Summer Science Fellows, NS from the St. Olaf CURI fund.

References

1. S. Moseley *et al.*, Phys. Rev. D **100**, 064031 (2019).
2. See, for example, N. Beverini *et al.*, J. Phys. Conf. Ser. **723**, 012061 (2016); A.D.V. Di Virgilio, these proceedings.
3. See, for example, D. Savoie *et al.*, Sci. Adv. **4**, eaau7948 (2018).
4. D. Colladay and V.A. Kostelecký, Phys. Rev. D **58**, 116002 (1998); V.A. Kostelecký, Phys. Rev. D **69**, 105009 (2004).
5. For a review, see J.D. Tasson, Rept. Prog. Phys. **77**, 062901 (2014).
6. Q.G. Bailey and V.A. Kostelecký, Phys. Rev. D **74**, 045001 (2006).
7. Q.G. Bailey and D. Havert, Phys. Rev. D **96**, 064035 (2017).
8. F. Bosi *et al.*, Phys. Rev. D **84**, 122002 (2011).
9. N. Scaramuzza and J.D. Tasson, in V.A. Kostelecký, ed., *CPT and Lorentz Symmetry VII*, World Scientific, Singapore 2017.
10. L. Shao and Q.G. Bailey, Phys. Rev. D **98**, 084049 (2018).
11. A. Ortolan *et al.*, J. Phys. Conf. Ser. **718**, 072003 (2016); A.D.V. Di Virgilio *et al.*, Eur. Phys. J. Plus **132**, 157 (2017).
12. *Data Tables for Lorentz and CPT Violation*, V.A. Kostelecký and N. Russell, 2019 edition, arXiv:0801.0287v12.

Lorentz Violation and Partons

Nathan Sherrill

Physics Department, Indiana University, Bloomington, IN 47405, USA

A parton-model description of high-energy hadronic interactions in the presence of Lorentz violation is presented. This approach is used to study lepton–hadron and hadron–hadron interactions at large momentum transfer. Cross sections for deep inelastic scattering and the Drell–Yan process are calculated at first order for minimal and nonminimal Lorentz violation. Estimated bounds are placed using existing LHC and future US-based Electron–Ion Collider data.

1. Introduction

The effective-field-theory framework to search for potential signals of Lorentz and CPT violation is known as the Standard-Model Extension (SME).[1,2] Despite numerous bounds that have been placed,[3] the QCD sector of the SME is relatively unexplored both experimentally and phenomenologically. This work, which is based on Ref. 4, describes a method for accessing some of this terrain from high-energy hadronic processes.

2. Description

In the presence of spin-independent Lorentz-violating effects on quarks, the dispersion relation is modified from the conventional one $k^2 = m^2$ and reads[5]

$$k^2 - F\left(k^\mu, m, \hat{\mathcal{Q}}\right) \equiv \widetilde{k}^2 = m^2. \tag{1}$$

In general, the function F depends on the coefficients for Lorentz violation $\hat{\mathcal{Q}}$, which are associated with operators of arbitrary mass dimension. To lowest order in electroweak interactions, partons (quarks) of momentum k^μ may be approximated as on-shell and massless, $k^2 = 0$, leading in the conventional case to the parameterization of the parton momentum as a fraction of the hadron momentum, $k^\mu = \xi p^\mu$. This parameterization, however, is inconsistent with Eq. (1); instead, the choice $\widetilde{k}^\mu = \xi p^\mu$ satisfies the conditions of interest. The implications of this parameterization are studied for minimal c-type and nonminimal $a^{(5)}$-type quark coefficients, which were first studied in Refs. 6,7, respectively, for the process of deep inelastic scattering (DIS).

2.1. *Deep inelastic scattering*

The DIS process $l + H \rightarrow l' + X$ describes a lepton l delivering a large momentum transfer $-q^2$ upon scattering from a hadron H producing a hadronic final state X. In the DIS limit $-q^2 \rightarrow \infty$ with $x = -q^2/(2p \cdot q)$ fixed, where p is the hadron momentum, the on-shell and massless limit of Eq. (1) yields the differential cross section

$$\frac{d\sigma}{dx dy d\phi} = \frac{\alpha^2 y}{2q^4} \sum_f e_f^2 \frac{1}{-\tilde{q}_f^2} L_{\mu\nu} H_f^{\mu\nu} f_f(\tilde{x}_f, c_f^{pp}), \qquad (2)$$

where

$$L_{\mu\nu} H_f^{\mu\nu} = 8 \left[2(\hat{k}_f \cdot l)(\hat{k}_f \cdot l') + \hat{k}_f \cdot (l - l')(l \cdot l') - 2(l \cdot l') c_f^{\hat{k}_f \hat{k}_f} \right.$$
$$+ 2(\hat{k}_f \cdot l) \left(c_f^{\hat{k}_f l'} + c_f^{l' \hat{k}_f} - c_f^{l'l'} \right) + 2(\hat{k}_f \cdot l') \left(c_f^{\hat{k}_f l} + c_f^{l \hat{k}_f} + c_f^{ll} \right) \right], \qquad (3)$$

with $\hat{k}_f^\mu \equiv \tilde{x}_f(p^\mu - c_f^{\mu p})$, $q_f^\mu = (\eta^{\mu\nu} + c_f^{\mu\nu})q_\nu$, and $y = (p \cdot q)/(p \cdot k)$. The shifted Bjorken variable $x_f = x \left(1 + 2c_f^{qq}/q^2 \right) + (x^2/q^2) \left(c_f^{pq} + c_f^{qp} \right)$ to first order in $c_f^{\mu\nu}$. The cross section for the quark $a^{(5)}$-type coefficients is given by Eq. (61) in Ref. 7, which is also consistent with Eq. (1) with the proton coefficients $a_p^{(5)\mu\alpha\beta} = 0$. Results for both cases are also consistent with the electromagnetic Ward identity and the operator product expansion.

2.2. *The Drell–Yan process*

Applying the parameterization Eq. (1) to the Drell–Yan process $H_1 + H_2 \rightarrow l_1 + l_2 + X$ gives the total cross section in the center-of-mass frame for the c-type coefficients:

$$\sigma = \frac{2\alpha^2}{3s} \frac{1}{Q^4} \int d\Omega_l \frac{d\xi_1}{\xi_1} \frac{d\xi_2}{\xi_2} \sum_f e_f^2 \left[(\tilde{k}_1 \cdot l_1)(\tilde{k}_2 \cdot l_2) + (\tilde{k}_1 \cdot l_2)(\tilde{k}_2 \cdot l_1) \right.$$

$$+ (\tilde{k}_1 \cdot l_1) \left(c_f^{\tilde{k}_2 l_2} + c_f^{l_2 \tilde{k}_2} \right) + (\tilde{k}_1 \cdot l_2) \left(c_f^{\tilde{k}_2 l_1} + c_f^{l_1 \tilde{k}_2} \right) + (\tilde{k}_2 \cdot l_1) \left(c_f^{\tilde{k}_1 l_2} + c_f^{l_2 \tilde{k}_1} \right)$$

$$+ (\tilde{k}_2 \cdot l_2) \left(c_f^{\tilde{k}_1 l_1} + c_f^{l_1 \tilde{k}_1} \right) - (\tilde{k}_1 \cdot \tilde{k}_2) \left(c_f^{l_1 l_2} + c_f^{l_2 l_1} \right) - (l_1 \cdot l_2) \left(c_f^{\tilde{k}_1 \tilde{k}_2} + c_f^{\tilde{k}_2 \tilde{k}_1} \right) \right]$$

$$\times \left(f_f(\xi_1, c_f^{p_1 p_1}) f_{\bar{f}}(\xi_2, c_f^{p_2 p_2}) + f_f(\xi_2, c_f^{p_2 p_2}) f_{\bar{f}}(\xi_1, c_f^{p_1 p_1}) \right). \qquad (4)$$

where $\tilde{k}_i^\mu = \xi_i p_i^\mu$ for $i = 1, 2$. As with DIS, the $a^{(5)}$-type quark coefficients yield a similar expression. The Ward identity is also satisfied in both cases.

3. Results

Using existing data from the LHC[8] and pseudodata for the future Electron–Ion Collider (EIC),[9] the best estimated limits for the equivalent coefficient combinations in the Sun-centered celestial-equatorial frame are shown in Table 1. These results suggest the c-type coefficients are more sensitive to

Table 1: Comparison of u-quark coefficients between the EIC and LHC. Bounds are reported in units of 10^{-5} and $10^{-6}\,\mathrm{GeV}^{-1}$ for minimal and nonminimal coefficients, respectively.

	EIC	LHC
$\lvert c_u^{XX} - c_u^{YY}\rvert$	0.74	15
$\lvert c_u^{XY}\rvert$	0.26	2.7
$\lvert c_u^{XZ}\rvert$	0.23	7.3
$\lvert c_u^{YZ}\rvert$	0.23	7.1
$\lvert a_{Su}^{(5)TXX} - a_{Su}^{(5)TYY}\rvert$	$0.15\,\mathrm{GeV}^{-1}$	$0.015\,\mathrm{GeV}^{-1}$
$\lvert a_{Su}^{(5)TXY}\rvert$	$0.12\,\mathrm{GeV}^{-1}$	$0.0027\,\mathrm{GeV}^{-1}$
$\lvert a_{Su}^{(5)TXZ}\rvert$	$0.13\,\mathrm{GeV}^{-1}$	$0.0072\,\mathrm{GeV}^{-1}$
$\lvert a_{Su}^{(5)TYZ}\rvert$	$0.13\,\mathrm{GeV}^{-1}$	$0.0070\,\mathrm{GeV}^{-1}$

lepton–hadron colliders, whereas $a^{(5)}$-type coefficients are more sensitive to hadron–hadron colliders.

Acknowledgments

This work was supported by the US Department of Energy under grant No. DE-SC0010120, the Indiana University Space Grant Consortium, and by the Indiana University Center for Spacetime Symmetries.

References

1. D. Colladay and V.A. Kostelecký, Phys. Rev. D **55**, 6760 (1997); Phys. Rev. D **58**, 116002 (1998).
2. V.A. Kostelecký, Phys. Rev. D **69**, 105009 (2004).
3. *Data Tables for Lorentz and CPT Violation,* V.A. Kostelecký and N. Russell, Rev. Mod. Phys. **83**, 11 (2011); 2019 edition arXiv:0801.0287.
4. V.A. Kostelecký, E. Lunghi, N. Sherrill, and A.R. Vieira, arXiv:1911.04002.
5. V.A. Kostelecký and M. Mewes, Phys. Rev. D **88**, 096006 (2013).
6. V.A. Kostelecký, E. Lunghi, and A.R. Vieira, Phys. Lett. B **769**, 272 (2017).
7. V.A. Kostelecký and Z. Li, Phys. Rev. D **99**, 056016 (2019).
8. CMS Collaboration, A.M. Sirunyan *et al.*, arXiv:1812.10529.
9. E. Lunghi and N. Sherrill, Phys. Rev. D **98**, 115018 (2018).

The Standard-Model Extension as an Effective Field Theory for Weyl Semimetals

Navin McGinnis

Physics Department, Indiana University, Bloomington, IN 47405, USA

Recent progress in the condensed-matter literature has revealed that novel features of materials with Weyl fermions can be modeled by an action analogous to Lorentz-violating QED. This suggests an application of the Standard-Model Extension to certain condensed-matter systems, in particular Weyl semimetals. We propose a systematic study of Fermi surfaces and transport phenomena to generalize these observations.

1. Introduction

The Standard-Model Extension (SME)[1] is a comprehensive framework developed to study observable effects of Lorentz- and CPT-symmetry breaking in particle physics and gravity. These effects are characterized in an effective field theory by extending Lorentz-invariant lagrangians by background fields which are not invariant under particle Lorentz transformations. The effective field theory for Lorentz-violating quantum electrodynamics (LVQED) is, for example, given by Eqs. (1) and (2) in Ref. 2.

Recently, it has been observed in the condensed-matter literature that hamiltonians describing Weyl semimetals exhibit similar features. However, from this microscopic approach only effects related to the b term in LVQED have been explored.[3] We reverse this perspective treating the action as fundamental. This allows us to generalize these observations and explore the physics of other SME coefficients in the context of condensed-matter systems. This contribution to the CPT'19 proceedings is based on Ref. 4.

2. Hamiltonians for Weyl semimetals and the SME

In Ref. 3 it was argued that the hamiltonian describing a Weyl semimetal on a lattice can be derived from the following continuum action:

$$S = \int d^4x \, \bar{\psi} \left[\gamma^\mu (i\partial_\mu - A_\mu - b_\mu \gamma^5) - m \right] \psi. \tag{1}$$

Further, the corresponding hamiltonian can be obtained simply by introducing a lattice regularization in momentum space, $k_i \to \sin k_i$, and $m \to m + t \sum_i (1 - \cos k_i)$, where t is the hopping amplitude, and the lattice spacing and Fermi velocity in the medium have been set to unity. Applying this procedure to Eq. (1) gives

$$
\mathcal{H} = [\sin(k_y)\sigma_x - \sin(k_x)\sigma_y] \otimes \tau_z + \sin(k_z)\mathbb{1} \otimes \tau_y + m\mathbb{1} \otimes \tau_x
$$
$$
+ t \sum_i [1 - \cos(k_i)] \, \mathbb{1} \otimes \tau_x + u^\mu b_\mu, \tag{2}
$$

where σ_i (τ_i) are the spin (orbital) Pauli matrices, and

$$
u^\mu = (\sigma_z \otimes \tau_y, -\sigma_x \otimes \tau_x, -\sigma_y \otimes \tau_x, \sigma_z \otimes \mathbb{1}). \tag{3}
$$

The last term is crucial for modeling the band structure of a Weyl semimetal. The b term sets a physical scale for the separation of Dirac nodes in the Fermi surface.

This method can be applied further to other coefficients in the SME. The H coefficient in LVQED is of particular interest since it also has mass dimension one, and thus also sets a physical scale in the Fermi surface. The contribution to the hamiltonian from the H term, in the basis where

$$
H^{\mu\nu} = \begin{pmatrix} 0 & \vec{H}^T \\ \hline -\vec{H} & \epsilon^{ijk} h^k \end{pmatrix}, \tag{4}
$$

is

$$
\mathcal{H} \supset \vec{H} \cdot \vec{\sigma}_H - \vec{h} \cdot \vec{\sigma}_h, \quad \vec{\sigma}_H = \begin{pmatrix} -\sigma_y \otimes \tau_y \\ \sigma_x \otimes \tau_y \\ -\mathbb{1} \otimes \tau_z \end{pmatrix}, \quad \vec{\sigma}_h = \begin{pmatrix} \sigma_x \otimes \mathbb{1} \\ \sigma_y \otimes \mathbb{1} \\ -\sigma_z \otimes \tau_x \end{pmatrix}. \tag{5}
$$

This result can be used to study the surface spectra using the method of self-adjoint operators.[5] We show the details of this in a forthcoming work.[4]

3. Transport

Another striking feature of Weyl semimetals is their unusual electromagnetic response.[6] It is known that the massless limit of LVQED with only the b term contains a chiral anomaly. The induced Chern–Simons action then leads to a current with nonzero divergence[7]

$$
\vec{j} = \frac{e^2}{2\pi^2} \left(\vec{b} \times \vec{E} - b_0 \vec{B} \right). \tag{6}
$$

The first term gives rise to the anomalous quantum Hall effect and the second the chiral magnetic effect. This result has attracted much attention in the condensed-matter community as it offers a field-theoretic understanding for the breaking of classical symmetries by quantum fluctuations in Weyl semimetals.

The subject of transport phenomena for other coefficients can be approached by considering the linear response of the system to a *background* electromagnetic field, $A_\nu^B(x)$. In this context, the induced current is given by

$$\langle j^\mu \rangle = \Pi^{\mu\nu} A_\nu^B, \qquad (7)$$

where $\Pi^{\mu\nu}$ is the one-loop vacuum polarization. Thus, the contribution from a given coefficient in the SME to the induced current can be included by calculating $\Pi^{\mu\nu}$ to any order in that coefficient. This formula for the linear response implies a result for the bulk conductivity of the material. From Eq. (7), it can be seen that any term in the vacuum polarization linear in the external momentum gives an induced current of the form $\langle j^\mu \rangle \sim (\sigma \cdot p)^{\mu\nu} A_\nu^B$. In position space, this gives an analogous formula to Ohm's law, $\langle j^\mu \rangle \sim (\sigma \cdot \partial)^{\mu\nu} A_\nu^B$.

Acknowledgments

This work was supported in part by the US Department of Energy, the National Science Foundation, the Brazilian agencies CNPq and FAPEMA, and the Indiana University Center for Spacetime Symmetries.

References

1. D. Colladay and V.A. Kostelecký, Phys. Rev. D **55**, 6760 (1997); Phys. Rev. D **58**, 116002 (1998); V.A. Kostelecký, Phys. Rev. D **69**, 105009 (2004).
2. V.A. Kostelecký, C.D. Lane, and A.G.M. Pickering, Phys. Rev. D **65**, 056006 (2002).
3. J. Behrends, S. Roy, M.H. Kolodrubetz, J.H. Bardarson, and A.G. Grushin, Phys. Rev. B **99**, 140201 (2019).
4. V.A. Kostelecký, R. Lehnert, N. McGinnis, M. Schreck, and B. Seradjeh, in preparation.
5. B. Seradjeh and M. Vennettilli, Phys. Rev. B **97**, 075132 (2018); A.T. Mostafa, G. Ortiz, and B. Seradjeh, Am. J. Phys. **84**, 858 (2016).
6. See, e.g., O.J. Franca, L.F. Urrutia, and O. Rodríguez-Tzompatzi, these proceedings.
7. M.M. Vazifeh and M. Franz, Phys. Rev. Lett. **111**, 027201 (2013).

Spacetime Nonmetricity and Lorentz Violation

Rui Xu

*Kavli Institute for Astronomy and Astrophysics, Peking University,
Beijing 100871, China*

Nonmetricity exists in gravitational theories when the connection is not metric compatible. To constrain it experimentally we construct general couplings between nonmetricity and a Dirac fermion in effective field theory. With the assumption of a nonzero background nonmetricity, the theory resembles the fermion sector of the Standard-Model Extension. By carefully choosing experiments that test Lorentz symmetry, their constraints on Lorentz violation are translated to the first experimental bounds on nonmetricity.

1. Metric-affine gravitational theories

Describing gravity as spacetime geometry is an elegant as well as successful idea initiated by Einstein's General Relativity. This theory is at the same time the simplest one to realize this idea, as its connection is simply related to the metric by virtue of the torsion-free and metric-compatible conditions. More general geometric theories of gravity that abandon either or both of the conditions to include torsion $T^\alpha{}_{\mu\nu} \equiv \Gamma^\alpha{}_{\mu\nu} - \Gamma^\alpha{}_{\nu\mu}$, or nonmetricity $N_{\mu\alpha\beta} \equiv D_\mu g_{\alpha\beta}$, or both are called metric-affine gravitation theories.[1]

Torsion has been studied for a long time, and various experiments have put stringent bounds on it.[2] On the other hand, while the geometric effect that the inner product of two vectors is no longer conserved in parallel transport when nonmetricity is present is known, potential physical effects of nonmetricity have not been studied previously. We show in this work that a nonzero background of nonmetricity causes Lorentz violation and can therefore be tested in high-precision modern experiments.

2. Couplings between nonmetricity and a single fermion

In effective field theory, the Lagrange density of a Dirac fermion with all possible couplings between nonmetricity and the fermion can be expressed as[3]

$$\mathcal{L} = \mathcal{L}_0 + \mathcal{L}_N^{(4)} + \mathcal{L}_N^{(5)} + \mathcal{L}_N^{(6)} + \ldots, \tag{1}$$

where $\mathcal{L}_0$ is the conventional Lagrange density for a Dirac fermion and $\mathcal{L}_N^{(d)}$ consists of all possible couplings between nonmetricity and fermion bilinear operators of mass dimension d. The explicit expressions for $\mathcal{L}_N^{(d)}$ are shown in Ref. 3. Generally, terms in $\mathcal{L}_N^{(d)}$ have $d-4$ derivatives acting on the fermion field generating roughly a factor of E^{d-4}, where E is the energy of the fermion. Therefore, those terms are suppressed by a factor of $\left(\frac{E}{E_{\max}}\right)^{d-4}$ with $E_{\max}$ being the energy limit below which the effective field theory is valid.

Following this reasoning, the Lagrange density (1), truncated at $d=7$, generates a modified Dirac equation including all the leading nonmetricity couplings. To experimentally test nonmetricity in a model-independent way, it is practically useful to approximate nonmetricity as a constant background in the spacetime regime of laboratory experiments. This leads to a correspondence between Eq. (1) and the Lagrange density in the fermion sector of the Standard-Model Extension,[4] indicating a violation of Lorentz symmetry when a nonmetricity background is present.

3. Experimental constraints on nonmetricity

The correspondence allows us to take advantage of the abundant results established in the Standard-Model Extension to investigate nonmetricity, especially using up-to-date experimental constraints on Lorentz violation to set limits on nonmetricity. A careful selection of the constraints on Lorentz violation in vacuum shows an upper limit of 10^{-22} GeV to 10^{-33} GeV for different components of nonmetricity.[3] Note that the results come from various sources including experiments with He–Xe dual masers,[5] experiments with Hg–Cs comagnetometer,[6] the muon $g-2$ experiment,[7] experiments measuring the hydrogen hyperfine transition,[8] and astrophysical observations of cosmic-ray Čerenkov radiation.[4,9]

Experimental tests of Lorentz violation in matter are relatively rare. A search for neutron spin rotation at the National Institute of Standards and Technology (NIST) Center for Neutron Research[10] sets an upper limit of 10^{-22} GeV on background nonmetricity in liquid ^{4}He.[11] This is the first and only constraint on nonmetricity in matter thus far.

Acknowledgments

This work was supported by the US Department of Energy under grant No. DE-SC0010120, by the US National Science Foundation under grant

No. PHY-1614545, by the Indiana University Center for Spacetime Symmetries, by the Indiana University Collaborative Research and Creative Activity Fund of the Office of the Vice President for Research, by the Indiana University Collaborative Research Grants program, by the National Science Foundation of China under grant No. 11605056, and by the Chinese Scholarship Council.

References

1. See, e.g., M. Blagojević and F.W. Hehl, eds., *Gauge Theories of Gravitation*, Imperial College Press, London, 2013.
2. V.A. Kostelecký, N. Russell, and J.D. Tasson, Phys. Rev. Lett. **100**, 111102 (2008); B.R. Heckel, E.G. Adelberger, C.E. Cramer, T.S. Cook, S. Schlamminger, and U. Schmidt, Phys. Rev. D **78**, 092006 (2008); R. Lehnert, W.M. Snow, and H. Yan, Phys. Lett. B **730**, 353 (2014); Y.N. Obukhov, A.J. Silenko, and O.V. Teryaev, Phys. Rev. D **90**, 124068 (2014); L.C. Garcia de Andrade, Mod. Phys. Lett. A **29**, 1450171 (2014).
3. J. Foster, V.A. Kostelecký, and R. Xu, Phys. Rev. D **95**, 084033 (2017).
4. V.A. Kostelecký and M. Mewes, Phys. Rev. D **88**, 096006 (2013).
5. F. Canè, D. Bear, D.F. Phillips, M.S. Rosen, C.L. Smallwood, R.E. Stoner, R.L. Walsworth, and V.A. Kostelecký, Phys. Rev. Lett. **93**, 230801 (2004); F. Allmendinger, W. Heil, S. Karpuk, W. Kilian, A. Scharth, U. Schmidt, A. Schnabel, Yu. Sobolev, and K. Tullney, Phys. Rev. Lett. **112**, 110801 (2014).
6. S.K. Peck, D.K. Kim, D. Stein, D. Orbaker, A. Foss, M.T. Hummon, and L.R. Hunter, Phys. Rev. A **86**, 012109 (2012).
7. G.W. Bennett *et al.*, Phys. Rev. Lett. **100**, 091602 (2008).
8. M.A. Humphrey, D.F. Phillips, and R.L. Walsworth, Phys. Rev. A **62**, 063405 (2000); D.F. Phillips, M.A. Humphrey, E.M. Mattison, R.E. Stoner, R.F.C. Vessot, and R.L. Walsworth, Phys. Rev. D **63**, 111101(R) (2001); M.A. Humphrey, D.F. Phillips, E.M. Mattison, R.F.C. Vessot, R.E. Stoner, and R.L. Walsworth, Phys. Rev. A **68**, 063807 (2003).
9. O. Gagnon and G.D. Moore, Phys. Rev. D **70**, 065002 (2004).
10. J.S. Nico, *et al.*, J. Res. NIST **110**, 137 (2005).
11. R. Lehnert, W.M. Snow, Z. Xiao, and R. Xu, Phys. Lett. B **772**, 865 (2017).

Lorentz-Violation Signal in the Comparison of Atomic Clocks

Xiao-yu Lu, Yu-Jie Tan, and Cheng-Gang Shao

MOE Key Laboratory of Fundamental Physical Quantities Measurements,
Hubei Key Laboratory of Gravitation and Quantum Physics,
PGMF and School of Physics, Huazhong University of Science and Technology,
Wuhan 430074, China

Lorentz-violation signals in the comparison of atomic clocks are analyzed in the Robertson–Mansouri–Sexl kinematical framework. As this framework describes deviations of the coordinate transformations from the Lorentz transformations from the viewpoint of the transformation modifications of time and space, the Lorentz-violating effects in atomic-clock comparisons have two sources: a time-delay effect $\alpha \frac{v^2}{c^2}$ and a structure effect $-\frac{\beta+2\delta}{3}\frac{v^2}{c^2}$. The Standard-Model Extension is a widely used dynamical framework to characterize Lorentz violation, in which a space-orientation dependence caused by a background field is regarded as the essential reason for Lorentz violation. Compared with the Robertson–Mansouri–Sexl framework, which only governs the kinematical properties via coordinate transformations, this dynamical framework provides a more complete and clear description of Lorentz violation.

1. Introduction

Lorentz invariance (LI) is a fundamental symmetry of spacetime, which postulates that experimental results are independent of the orientation and state of uniform motion of the apparatus.[1] As LI is at the foundation of both the Standard Model of particles physics and General Relativity, its study is an important subject in the physical sciences. Here, we investigate the Lorentz-violation (LV) effects in the Robertson–Mansouri–Sexl (RMS) framework,[2,3] and perform a simple comparison with the Standard-Model Extension (SME) framework.[4,5] The RMS framework considers the speed of light to be anisotropic, and it also postulates that there is a preferred universal frame in which light propagates conventionally as measured using a set of rods and clocks. These RMS rods and clocks are isotropic and the photon is anisotropic, while for SME rods and clocks the case can be opposite. Since LV may result in corrections to transition frequencies, atomic-clock comparisons are excellent tools to search for LV effects, and we focus here on analyzing this particular situation.

2. Lorentz violation in atomic-clock comparisons

The violation of LI is described in the RMS kinematical framework as the deformation of Lorentz transformations, and it postulates the existence of a preferred frame Σ, that is usually taken to be the cosmic microwave background frame. If the laboratory reference frame S has speed v with respect to Σ, the transformation between these two frames can be written as[2,3]

$$t = aT + \vec{\varepsilon} \cdot \vec{x}, \quad x = b(X - vT), \quad y = dY, \quad z = dZ, \tag{1}$$

with $a(v) = 1 + (\alpha - \frac{1}{2})\frac{v^2}{c^2}$, $b(v) = 1 + (\beta + \frac{1}{2})\frac{v^2}{c^2}$, and $d(v) = 1 + \delta\frac{v^2}{c^2}$,[6] which contains the Lorentz transformations as the special case $\alpha = \beta = \delta = 0$. For the comparison of clock frequencies, the violation of LI in frame S can be detected through measuring the anisotropy of the speed of light. Analyzing light-clock and atomic-clock comparisons in Fig. 1a and 1b,[7,8] the LV signal of clock-comparison experiments is

$$\Delta_{LV} = \alpha\frac{v^2}{c^2} - \frac{\beta + 2\delta}{3}\frac{v^2}{c^2}, \tag{2}$$

where the first term characterizes a time-delay effect, and the second one a structure effect.[9]

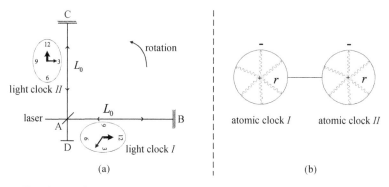

Fig. 1: Two kinds of clock-comparison experiments. Part (a) shows the comparison of light clocks: this is similar to a Michelson interferometer with both arms of the same length L_0, where each interferometer arm can be considered as a light clock. Part (b) depicts the comparison of atomic clocks: two atomic clocks are located in different places.

The SME provides a general theoretical framework for studying LV, such as the violation in the photon, matter, and gravity sectors, etc. Compared with RMS framework, the SME provides a vast parameter space. In this framework, LV background fields are postulated, and different coordinate systems are linked by the usual Lorentz transformations. A special case of the SME can be matched to the RMS framework. In this case, the photon sector of the SME is LI, and the main LV effects are embodied in the matter sector.[10] The RMS formalism can thus be regarded as a special limit of the SME.

3. Conclusion

Based on the analysis of LI violations for light-clock comparisons in RMS framework, we studied LV effects in atomic-clock comparisons, which include a time-delay and a structure effect. In addition, we also provided a simple explanation for the different signals of LI violation for atomic-clock comparisons in the RMS and SME frameworks. For the RMS framework, the LV effects arise from the deformation of the Lorentz transformation, while for the SME, they originate in the nondynamical background fields.

Acknowledgments

This work is supported by the Postdoctoral Science Foundation of China under grant Nos. 2017M620308 and 2018T110750.

References

1. O. Bertolami and J. Páramos, in A. Ashtekar and V. Petkov, eds., *Springer Handbook of Spacetime*, Springer, Berlin, 2014.
2. H.P. Robertson, Rev. Mod. Phys. **21**, 378 (1949).
3. R. Mansouri and R.U. Sexl, Gen. Rel. Grav. **8**, 497 (1977).
4. D. Colladay and V.A. Kostelecký, Phys. Rev. D **55**, 6760 (1997); **58**, 116002 (1998).
5. V.A. Kostelecký, Phys. Rev. D **69**, 105009 (2004).
6. Y.Z. Zhang, Gen. Rel. Grav. **27**, 475 (1995).
7. C. Eisele, A.Y. Nevsky, and S. Schiller, Phys. Rev. Lett. **103**, 090401 (2009).
8. P. Delva *et al.*, Phys. Rev. Lett. **118**, 221102 (2017).
9. X.Y. Lu, Y.J. Wang, Y.J. Tan, and C.-G. Shao, Phys. Rev. D **98**, 096022 (2018).
10. V.A. Kostelecký and M. Mewes, Phys. Rev. D **80**, 015020 (2009).

Comparative Penning-Trap Tests of Lorentz and CPT Symmetry

Yunhua Ding

Physics Department, Northern Michigan University, Marquette, MI 49855, USA

The theoretical and experimental prospects for Lorentz- and CPT-violating quantum electrodynamics in Penning traps are reviewed in this work. With the recent reported results for the measurements of magnetic moments for both protons and antiprotons, improvements with factors of up to 3000 for the constraints of various coefficients for Lorentz and CPT violation are obtained.

1. Introduction

Among the most fundamental symmetries of relativity and particle physics are Lorentz and CPT invariance. However, in recent years it has been suggested that tiny violations of Lorentz and CPT symmetry are possible in a unified theory of gravity and quantum physics, such as strings.[1] The comprehensive and relativistic description of such violations is given by the Standard-Model Extension (SME),[2] a general framework constructed from General Relativity and the Standard Model by adding to the action all possible Lorentz-violating terms. Each of such terms is formed as the contraction of a Lorentz-violating operator with a corresponding coefficient controlling the size of Lorentz violation. High-precision experiments across a broad range of subfields of physics, including Penning traps, provide striking constraints on the coefficients for Lorentz violation.[3] In the context of minimal SME, with operators of mass dimension restricted to $d \leq 4$, several studies of observables for Lorentz violation in Penning traps have been conducted.[4] The relevant theory of Lorentz-violating electrodynamics with nonminimal operators of mass dimensions up to six was also developed.[5] More recently, this treatment was generalized to include operators of arbitrary mass dimension using gauge field theory.[6] In this work, we further the study of experimental observables for Lorentz violation by comparing different Penning-trap experiments and extract new constraints on various coefficients for Lorentz violation using the recently published results for magnetic moments from the BASE collaboration.[7,8]

2. Theory

For a charged Dirac fermion ψ with mass m confined in a Penning trap, the magnetic moment and the related g factor of the particle can be obtained by measuring two frequencies, the Larmor frequency $\nu_L = \omega_L/2\pi$ and the cylotron frequency $\nu_c = \omega_c/2\pi$, and determining their ratio $g/2 = \nu_L/\nu_c = \omega_L/\omega_c$. In an ideal Penning-trap experiment with the magnetic field $\boldsymbol{B} = B\hat{x}_3$ lying along the positive x^3 axis of the apparatus frame, the leading-order contributions from Lorentz violation to ω_c for both fermion and antifermions vanish, while the corrections to ω_L are given by

$$\delta\omega_L^w = 2\widetilde{b}_w^3 - 2\widetilde{b}_{F,w}^{33}B, \quad \delta\omega_L^{\overline{w}} = -2\widetilde{b}_w^{*3} + 2\widetilde{b}_{F,w}^{*33}B, \tag{1}$$

where $w = e^-, p$ for electrons and protons and $\overline{w} = e^+, \overline{p}$ for positrons and antiprotons. The tilde quantities are defined as a combination of different coefficients for Lorentz violation.[5]

The expressions for the shifts in the anomaly frequencies (1) are valid in the apparatus frame where the direction of the magnetic field is chosen to be in the positive $\hat{x}_3$ direction. The results in terms of the constant coefficients in the Sun-centered frame requires a transformation matrix between the two frames,[9] which reveals the dependence of the anomaly frequencies on the sidereal time and the geometric conditions of the experiment.

3. Applications

For a proton or antiproton confined in a Penning trap, the relevant experiment-independent observables for the study of magnetic moments are the 18 coefficients for Lorentz violation $\widetilde{b}_p^J$, $\widetilde{b}_p^{*J}$, $\widetilde{b}_{F,p}^{(JK)}$, and $\widetilde{b}_{F,p}^{*(JK)}$, where $J, K = X, Y, Z$ in the Sun-centered frame. A comparison of the magnetic moments of protons and antiprotons in different Penning-trap experiments offers an excellent opportunity to constrain various of the 18 coefficients for Lorentz violation listed above. An analysis involving the comparison of the magnetic moments of protons from the BASE collaboration at Mainz and of antiprotons from the ATRAP experiment at CERN is given in our previous work.[5] Recently, the sensitivities of the magnetic moments for both protons and antiprotons have been improved by several orders of magnitude by the BASE collaboration at Mainz and CERN.[7,8] Here, we conduct a similar comparative analysis using the recent published results to extract improved constraints on coefficients for Lorentz violation.

For the magnetic moment of the proton measured at Mainz, the laboratory colatitude is $\chi \simeq 40.0°$ and the applied magnetic field $B \simeq 1.9\,\mathrm{T}$

points to local south, which corresponds to the $\hat{x}$ direction in the standard laboratory frame. For the antiproton magnetic-moment measurement at CERN, the laboratory colatitude is $\chi^* \simeq 43.8°$ and the magnetic field $B^* \simeq 1.95\,\mathrm{T}$ points $\theta = 60.0°$ east of local north. The experimental data for both experiments were taken over an extended time period, so we can plausibly average the sidereal variations to be zero, leaving only the constant parts. Together with the numerical values of other quantities reported by both BASE measurements, we obtain bounds for $\widetilde{b}_p^Z$, $\widetilde{b}_p^{*Z}$, $\widetilde{b}_{F,p}^{(XX)}$, $\widetilde{b}_{F,p}^{*(XX)}$, $\widetilde{b}_{F,p}^{(YY)}$, $\widetilde{b}_{F,p}^{*(YY)}$, $\widetilde{b}_{F,p}^{(ZZ)}$, and $\widetilde{b}_{F,p}^{*(ZZ)}$, with improvement factors of up to 3000 compared to the previous results.[5]

Note that among the 18 independent observables in Penning-trap experiments a large number remain unexplored to date. Performing a study of sidereal variations could in principle provide sensitivities to other components of the tilde coefficients. Such analyses could become possible once the quantum-logic readout currently under development at the BASE collaboration becomes feasible.[10]

Acknowledgments

This work was supported in part by the US Department of Energy and by the Indiana University Center for Spacetime Symmetries.

References

1. V.A. Kostelecký and S. Samuel, Phys. Rev. D **39**, 683 (1989); V.A. Kostelecký and R. Potting, Nucl. Phys. B **359**, 545 (1991); Phys. Rev. D **51**, 3923 (1995).
2. D. Colladay and V.A. Kostelecký, Phys. Rev. D **55**, 6760 (1997); Phys. Rev. D **58**, 116002 (1998); V.A. Kostelecký, Phys. Rev. D **69**, 105009 (2004).
3. *Data Tables for Lorentz and CPT Violation,* V.A. Kostelecký and N. Russell, 2019 edition, arXiv:0801.0287v12.
4. R. Bluhm, V.A. Kostelecký, and N. Russell, Phys. Rev. Lett. **79**, 1432 (1997); Phys. Rev. D **57**, 3923 (1998).
5. Y. Ding and V.A. Kostelecký, Phys. Rev. D **94**, 056008 (2016).
6. V.A. Kostelecký and Z. Li, Phys. Rev. D **99**, 056016 (2019).
7. C. Smorra *et al.*, Nature **550**, 371 (2017).
8. G. Schneider *et al.*, Science **358**, 1081 (2017).
9. V.A. Kostelecký and M. Mewes, Phys. Rev. D **66**, 056005 (2002).
10. T. Meiners *et al.*, in V.A. Kostelecký, ed., *Proceedings of the Seventh Meeting on CPT and Lorentz Symmetry*, World Scientific, Singapore, 2017; M. Niemann *et al.*, in V.A. Kostelecký, ed., *Proceedings of the Sixth Meeting on CPT and Lorentz Symmetry*, World Scientific, Singapore, 2014.

CPT-violating Effects on Neutron Gravitational Bound States

Zhi Xiao

Department of Mathematics and Physics, North China Electric Power University, Beijing 102206, China

Analytical solutions with effective CPT-violating spin–gravity corrections to the neutron's gravitational bound states are obtained. The helicity-dependent phase evolution due to $\vec{\sigma} \cdot \vec{b}$ and $\vec{\sigma} \cdot \hat{\vec{p}}$ couplings not only leads to spin precession, but also to transition-frequency shifts between different gravitational bound states. Utilizing transition frequencies measured in the qBounce experiment,[1] we obtain the rough bound $|\vec{b}| < 6.9 \times 10^{-21}\,\mathrm{GeV}$. Incorporating known systematic errors may lead to more robust and tighter constraints.

1. Introduction

Lorentz-violating matter–gravity couplings[2] open a broad and interesting avenue for testing Lorentz symmetry. Recently, spin-independent Lorentz-violating neutron–gravity couplings have been thoroughly studied[3] in an attempt to analyze the GRANIT experiment.[4] However, to our best knowledge, an extensive study of spin-dependent fermion–gravity couplings is still under development.[5] Here, we provide a first glimpse at CPT-violating spin-dependent neutron–gravity couplings. A more detailed and complete analysis can be found in Ref. 6.

2. LV corrections due to spin-dependent interactions

The main vertical hamiltonian after averaging over the horizontal degrees of freedom is[3]

$$\hat{H} = \frac{\hat{p}_z^2}{2m_I} + m_G\, g\, z - \vec{\sigma} \cdot \vec{b}\,(1 + \Phi_0),\tag{1}$$

where we have started with the hamiltonian in Ref. 7 and performed a series of redefinitions and approximations. The stationary solution of (1) is

$$\Psi_\perp(t,z) = \frac{1}{\sqrt{2}} \begin{pmatrix} \left[\cos(\frac{\Omega}{2}t) + ir_+ \sin(\frac{\Omega}{2}t)\right] e^{-\frac{i\omega}{2}t} \\ \left[\cos(\frac{\Omega}{2}t) - ir_- \sin(\frac{\Omega}{2}t)\right] e^{+\frac{i\omega}{2}t} \end{pmatrix} \phi_n(z) e^{-iE_n t},\tag{2}$$

where $\phi_n(z) \equiv \dfrac{\text{Ai}[\frac{z}{L_c} - x_{n+1}]}{L_c^{1/2}|\text{Ai}'[-x_{n+1}]|}$, $\Omega \equiv \sqrt{\omega^2 + 4\omega B_0 \cos\theta + 4B_0^2}$, and $r_\pm \equiv \left[\omega + 2B_0(\cos\theta \pm \sin\theta e^{-i\phi})\right]/\Omega$; the initial state is assumed to be an eigenstate of σ_x, $|X\uparrow\rangle$.

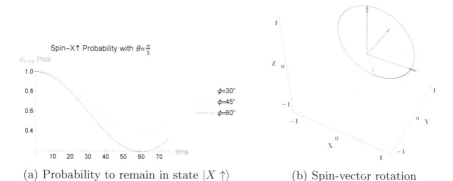

(a) Probability to remain in state $|X\uparrow\rangle$ (b) Spin-vector rotation

Fig. 1: For (a), $\omega = 0$, $\theta = \pi/3$, and $B_0 = \pi/(120\,\text{ms}) = 1.7 \times 10^{-23}$ GeV was chosen. The B_0 value lies outside the bound $|\vec{b}| < 10^{-29}$ GeV but was selected nevertheless to emphasize the probability variation. Note the ϕ dependence. Panel (b) shows the evolution of the spin vector for the initial state $|X\uparrow\rangle$. Again, a rather large $B_0 = \pi/(120\,\text{ms})$ was chosen. The other parameters are $\theta = \pi/6$, $\phi = \omega = 0$.

The probability profile for unrealistically large $B_0 \equiv |\vec{b}|$ as well as the spin-precession on the Bloch sphere are shown in Fig. 1. The eigensolution of $\hat{H} = \hat{p}_z^2/(2m_I) + m_G\, g\, z - \bar{b}_{\text{eff}}\,(1 + \Phi_0)\,\sigma_z \hat{p}_z/m_I$ is

$$\Psi(t, z) = \begin{pmatrix} c_1\, e^{i[\bar{b}_{\text{eff}}(1+\Phi_0)]z} \\ c_2\, e^{-i[\bar{b}_{\text{eff}}(1+\Phi_0)]z} \end{pmatrix} \text{Ai}[\tfrac{z}{L_c} - x_{n+1}] e^{-i\{E_n - \frac{[\bar{b}_{\text{eff}}(1+\Phi_0)]^2}{2m_I}\}t}, \quad (3)$$

where the parity-odd nature of $\vec{\sigma} \cdot \hat{\vec{p}}$ again dictates opposite phase evolutions for the two helicity components, and c_1 and c_2 are constants to be determined.

For $\hat{H}_{LV} = -\vec{\sigma} \cdot \vec{b}\,(1 + \Phi_0 + gz) - \bar{b}_{\text{eff}}\,(1 + \Phi_0)\,\sigma_z \hat{p}_z/m_I$, we can use the matrix elements

$$\begin{pmatrix} \langle n+|\hat{H}_{LV}|n, +\rangle & \langle n+|\hat{H}_{LV}|n, -\rangle \\ \langle n-|\hat{H}_{LV}|n, +\rangle & \langle n-|\hat{H}_{LV}|n, -\rangle \end{pmatrix}$$

$$= -B_0 \left[(1 + \Phi_0) + \tfrac{2}{3}g L_c x_{n+1}\right] \begin{pmatrix} \cos\theta & \sin\theta e^{-i\phi} \\ \sin\theta e^{i\phi} & -\cos\theta \end{pmatrix} \quad (4)$$

to calculate the shift in the eigenenergies

$$\delta E_n = \mp B_0 \left[(1 + \Phi_0) + \tfrac{2}{3} g L_c x_{n+1} \right], \tag{5}$$

where the upper or lower sign depends on whether the spin state is $|\hat{n}+\rangle = (e^{-i\phi}\cos\tfrac{1}{2}\theta, \sin\tfrac{1}{2}\theta)$ or $|\hat{n}-\rangle = (\sin\tfrac{1}{2}\theta, -e^{i\phi}\cos\tfrac{1}{2}\theta)$, respectively. From (5), the transition-frequency shift is given by $\delta\nu^\pi_{mn} = \mp\frac{2g}{3h}\,|\vec{b}|\,L_c(x_{m+1} - x_{n+1})$. Comparing $\delta\nu^\pi_{mn}$ with the precisely measured frequencies in the qBounce experiment,[1] we obtain the rough upper bound $|\vec{b}| < 6.946 \times 10^{-21}$ GeV.

3. Summary

In this work, we have discussed CPT-violating spin–gravity corrections on the neutron's gravitational bound states. With several analytical solutions, we have demonstrated that the phase evolution depends on the helicity of the wave-function components. The resulting phenomena are spin precession and θ- and ϕ-dependent probability variations. Using degenerate perturbation theory, we have also calculated the transition-frequency shift. Comparison with the measurements in Ref. 1 has yielded the rough bound $|\vec{b}| < 6.9 \times 10^{-21}$ GeV, which can be improved further if systematic errors from known physics are taken into account, or if polarized neutrons are used in the future.

Acknowledgments

The author appreciates valuable encouragement and helpful discussions with M. Snow and A. Kostelecký as well as help from many others. This work is partially supported by the National Science Foundation of China under grant Nos. 11605056, 11875127, and 11575060.

References

1. G. Cronenberg *et al.*, Nature Phys. **14**, 1022 (2018).
2. V.A. Kostelecký and J.D. Tasson, Phys. Rev. Lett. **102**, 010402 (2009); Phys. Rev. D **83**, 016013 (2011).
3. A. Martín-Ruiz and C.A. Escobar, Phys. Rev. D **97**, 095039 (2018).
4. V.V. Nesvizhevsky *et al.*, Nature **415**, 297 (2002).
5. V.A. Kostelecký and Z. Li, to appear. See also Z. Li, these proceedings.
6. Z. Xiao, arXiv:1906.00146 [hep-ph].
7. Y. Bonder, Phys. Rev. D **88**, 105011 (2013).

Lorentz Violation in the Matter–Gravity Sector

Zonghao Li

Physics Department, Indiana University, Bloomington, IN 47405, USA

We construct the general Lorentz-violating effective field theory in curved spacetime and the corresponding nonrelativistic Hamiltonian in the Earth's gravitational field. Applying this general framework to three types of experiments, free-dropping, interferometer, and bound-state experiments, we extract first constraints on certain new coefficients in the matter–gravity sector.

Lorentz symmetry is a fundamental symmetry in both General Relativity (GR) and the Standard Model (SM); it deserves to be precisely tested in experiments. Moreover, Lorentz violation has been a popular candidate in recent years as a low-energy remnant of the unification of GR and the SM. To study Lorentz violation, D. Colladay and V.A. Kostelecký developed a comprehensive framework, the Standard-Model Extension (SME), in the context of effective field theory.[1,2] The SME contains GR and the conventional SM coupled to GR, and it adds all possible Lorentz-violating modifications at the Lagrangian level. We have constructed the general Lorentz-violating terms in flat spacetime.[3] Expanding this method, we further built the general Lorentz-violating terms in curved spacetime.[4]

We first sketch the procedure of building the Lagrange density and then investigate its applications to experiments. We focus on the experiments in the matter–gravity sector, which has been a fruitful area in the search for Lorentz-violating signals.[5] Using the general framework we built, we can study some new effects, including those from nonminimal coefficients and spin–gravity couplings. The present contribution to the CPT'19 proceedings is based on results in Ref. 4.

In flat spacetime, the general Lorentz-violating effective field theory is built from gauge-covariant operators to preserve gauge invariance.[3] In curved spacetime, this idea is expanded to gauge-covariant spacetime-tensor operators to incorporate observer diffeomorphism invariance. Coupling those operators with coefficients for Lorentz violation, we construct the full SME in curved spacetime.

To make applications to experiments easier, we transfer the full Lagrange density into its nonrelativistic Hamiltonian. We first turn the full Lagrange density into its linearized Lagrange density by taking the linearized limit in a weak gravity field. Then, we extract the relativistic Hamiltonian by solving the equation of motion. Finally, we get the nonrelativistic Hamiltonian by the Foldy–Wouthuysen transformation,[6] which is a systematic procedure to extract nonrelativistic Hamiltonians from relativistic Hamiltonians for fermions.

The nonrelativistic Hamiltonian for a spin-1/2 fermion in the Earth's gravitational field can be expressed as:

$$H = H_0 + H_\phi + H_g + H_{\sigma\phi} + H_{\sigma g}, \tag{1}$$

where H_0 is the conventional term, and the remaining terms are corrections from the SME and depend on the coefficients for Lorentz violation. These corrections contain couplings between the gravitational field and the position, momentum, and spin of the particle. Terms in H_ϕ and H_g are independent of the spin, but those in $H_{\sigma\phi}$ and $H_{\sigma g}$ depend on the spin, which permits us to study spin–gravity couplings in the SME framework for the first time.

Three types of experiments, free-dropping, interferometer, and bound-state experiments, are analyzed as examples of the applications of the general framework in the matter–gravity sector.

Free-dropping experiments compared the accelerations of freely falling atoms with different inner structures. We focus on free drops of atoms with different spin polarizations to test spin–gravity couplings. An experiment in Italy dropped unpolarized ^{87}Sr atoms and tested the broadening of the accelerations due to different spin polarizations.[7] Another experiment in China dropped ^{87}Rb atoms with different spin polarizations and measured the difference between the accelerations of the atoms.[8] The relative discrepancies of accelerations are bounded around 10^{-7} in these experiments. These are transformed into bounds on certain coefficients for Lorentz violation and spin–gravity couplings.[4] Notice that different experiments constrain different coefficients for Lorentz violation; more experiments, such as dropping hydrogen and antihydrogen,[9] are needed for full coverage of those coefficients.

Interferometer experiments measured gravity-induced quantum phase shifts. Using the classic COW experiment performed by R. Colella, A.W. Overhauser, and S.A. Werner in 1975,[10] we get bounds on certain spin-independent coefficients.[4] More recent interferometer experiments[11] using magnetic fields to split neutron beams are expected to improve the result and constrain some spin-dependent coefficients.

Bound-state experiments measured the energies of the bound states of neutrons in the Earth's gravitational field. The original experiment measured the critical heights, [12] which are related to the energies by $mgz = E$. The improved version measured the transition frequencies between different energy states. [13] A sensitivity of 10^{-2} was attained and is used to constrain certain coefficients for Lorentz violation and spin–gravity couplings. [4]

In summary, fruitful results have been extracted from these three types of experiments. More experiments are needed for full coverage of the coefficients for Lorentz violation and spin–gravity couplings.

Acknowledgments

This work was supported in part by the US Department of Energy and by the Indiana University Center for Spacetime Symmetries.

References

1. D. Colladay and V.A. Kostelecký, Phys. Rev. D **55**, 6760 (1997); Phys. Rev. D **58**, 116002 (1998).
2. V.A. Kostelecký, Phys. Rev. D **69**, 105009 (2004).
3. V.A. Kostelecký and Z. Li, Phys. Rev. D **99**, 056016 (2019).
4. V.A. Kostelecký and Z. Li, in preparation.
5. V.A. Kostelecký and J. Tasson, Phys. Rev. D **83**, 016013 (2011).
6. L.L. Foldy and S.A. Wouthuysen, Phys. Rev. **78**, 29 (1950).
7. M.G. Tarallo *et al.*, Phys. Rev. Lett. **113**, 023005 (2014).
8. X.C. Duan *et al.*, Phys. Rev. Lett. **117**, 023001 (2016).
9. S. Aghion *et al.*, Nat. Commun. **5**, 4538 (2014); C. Amole *et al.*, Phys. Rev. Lett. 112, 121102 (2014); P. Indelicato *et al.*, Nat. Commun. **4**, 1787 (2013).
10. R. Colella, A.W. Overhauser, and S.A. Werner, Phys. Rev. Lett. **34**, 23 (1975).
11. V. de Haan *et al.*, Phys. Rev. A **89**, 063661 (2014).
12. V.V. Nesvizhevsky *et al.*, Nature **415**, 297 (2002).
13. G. Cronenberg *et al.*, Nat. Phys. **14**, 1022 (2018).

Author Index

CPSIA information can be obtained
at www.ICGtesting.com
Printed in the USA
BVHW040803140420
576745BV00007B/7

9 789811 213977